Human Nature in an Age of Biotechnology

Philosophy of Engineering and Technology

VOLUME 14

For further volumes:
http://www.springer.com/series/8657

Tamar Sharon

Human Nature in an Age of Biotechnology

The Case for Mediated Posthumanism

 Springer

Tamar Sharon
Philosophy
Maastricht University
Maastricht, Limburg, The Netherlands

ISSN 1879-7202 ISSN 1879-7210 (electronic)
ISBN 978-94-007-7553-4 ISBN 978-94-007-7554-1 (eBook)
DOI 10.1007/978-94-007-7554-1
Springer Dordrecht Heidelberg New York London

Library of Congress Control Number: 2013949769

Printed on acid-free paper

Springer is part of Springer Science+Business Media (www.springer.com)

To Lior Dangoor
my beloved ethicist of technology

Contents

Chapter 1
Introduction

The question of what it means to be human surfaces time and again in periods of important technological change. As if, once detached from the labor of their creation, technologies then take on the capacity of philosophical anthropologists: signaling to us, undeterred by their own non-humanness, that the fact of their existence solicits a clear definition of human nature. In our current technologized culture, where the life sciences themselves are increasingly merging with technology in the form of reproductive, genetic and neuro-technology, the question of what it means to be human has taken on a new urgency.

This is in great part prompted by what seems to be the sheer novelty of many of the emerging biotechnologies that surround us, from embryo selection to pre-implantation diagnostics, to the use of cloning techniques for reproductive and therapeutic purposes, to neural implants and mood-altering and memory-enhancing psychopharmaceuticals. The perceived novelty of emerging biotechnologies lies in the unprecedented degree of intervention into matter, life processes and nature that they offer, and the implications this has for notions like biological determinism, authenticity and even fate. It also lies in their ability to profoundly reconfigure, if not erase, clear and meaningful boundaries. Boundaries that have acted as stable and reliable frameworks for many of our traditional categories of thought, like those between humans and machines, between nature and technology, between treatment and enhancement, between the born and the made. Many of these distinctions are giving way to new entities and categories, such as technologically enhanced humans, non-organic life, intelligent machines and bio-engineered nature, that have customarily been the stuff of science fiction writers. Above all, the novel character of these technological developments lies in the idea that they are rendering the notion of "human nature" ever more uncertain, by both complicating the question of what it means to be "human" and by challenging the fixity of what is meant by "nature". This book emerged from the conviction that the dominant theoretical approaches concerned with the implications of new biotechnologies for what it means to be human are insufficient, and that a new perspective needs to be developed.

T. Sharon, *Human Nature in an Age of Biotechnology: The Case for Mediated Posthumanism*, Philosophy of Engineering and Technology 14, DOI 10.1007/978-94-007-7554-1_1, © Springer Science+Business Media Dordrecht 2014

1.1   A Polarized Framework for Discussion

Emerging bio- and enhancement technologies engender a range of reactions, extending from hope to uncertainty and from wonder to fear. For some, these developments hold the promise of liberating humans from the burdens of their biological, neurological and psychological determination, of introducing a measure of control into "nature's lottery." Thus assisted reproduction, from relatively simple techniques of artificial insemination to in vitro fertilization and other forms of reproductive interventions, seek to overcome natural limitations of age and infertility; the new generation of psychopharmaceuticals seeks to modify forms of emotion, cognition and perhaps personality that are seen as "hardwired" in the brain; and the new genetics seeks to select against certain hereditary diseases already in the embryonic context, or to offer a personalized medicine tailored to each and every individual's genotype in the clinical context. For others, more critical of emerging biotechnologies, these developments conjure up a world of engineered humans, designed on demand, a world in which a rampant technological "hubris" might lead us straight into a very inhuman future.

It is in this polarized framework of celebration and condemnation that public debate concerning the ethical, legal and social implications of emerging biotechnologies is most commonly organized, epitomized by the approaches of what are known as *transhumanism* and *bioconservatism*. Generally speaking, transhumanists, a group which includes theorists and futurists such as Nick Bostrom, Julian Savulescu, James Hughes, Ray Kurzweil and Hans Moravec, argue that the human condition should be improved via the use of new technologies where this is possible. Bioconservatives, as they are often disdainfully called by their opponents, among them prominent political philosophers and bioethicists like Francis Fukuyama, Leon Kass, George Annas and Michael Sandel, are very skeptic about technological transformations of the living world, and argue for a strict regulation of new biotechnologies. The widespread use of emerging biotechnologies, these theorists caution, at least for enhancement purposes, may introduce new forms of inequality and discrimination and, perhaps more importantly, violate a fundamental human essence.

In its immediate form, this polarized debate is usually framed in terms of risk and of access to the technologies, with concerns of social justice as a common backdrop. Bioconservative arguments revolve around the need for precautionary measures in assessing the long-term effects of biotechnologies; issues of discrimination arising from the unequal access to new biotechnologies that might turn financial disadvantages into biological ones (Fukuyama 2002; McKibben 2003); and conformism (Sandel 2007) or eugenic concerns (Fukuyama 2002; Habermas 2003) in the context of cognitive enhancements and preimplantation genetic diagnosis. Conversely, transhumanist claims contest that the precautionary principle stifles technological progress (Bostrom 2002); that questions of access should be dealt with by making the technologies widely available, if needed via compensating social policies (Hughes 2004); that enhancement technologies can actually *alleviate* inequalities that arise from the unequal distribution of biological capacities at birth (Hughes 2009);

and, pushing the market logic to its extreme, that a "liberal eugenics" will actually help express the diverse and particular values of individuals rather than narrow them down (Agar 2004; Savulescu 2001).

If the debate on emerging bio- and enhancement technologies is then usually articulated within a framework of risks, access, and social justice – relatively commensurable terms – this tends to obscure the more difficult aspects of the debate as a dispute about what it means to be human. For both bioconservatives and transhumanists, though this may be less obvious for the latter than for the former, what is at stake here is human nature. Thus the bioconservative critique of emerging bio- and enhancement technologies proceeds from the idea that technological intervention, at least for enhancement purposes, poses a threat to human nature and the values and virtues that humans have developed as a result of the necessity to deal with the imperfection inherent to this nature. The *givenness* of human nature can be seen as the defining characteristic, or fundamental essence, of what it means to be human in this perspective. For transhumanists, as a species humans have always struggled to expand their capacities in ways that humans before them were not able to, and emerging biotechnologies, be they intended for therapeutic or enhancement purposes, are the most recent expression and instrument of this essentially human drive towards self-improvement. This transcendental aspiration, or *transformative* essence, can be seen as the defining characteristic of what it means to be human in this perspective.

These clashing accounts of human nature are not as antithetical as they may seem, however. Essentially, as will be argued in more detail, they are both grounded in the humanist narrative of the human as an autonomous, unique and fixed entity, that is separate from its environment in a distinct way. In other words, these seemingly conflicting views of human nature are two versions of the humanist worldview that posits a foundational ontological divide between humans and the rest world – where the "rest of the world" takes on different attributes in different contexts, as "objects", as the "artifactual", as "non-humans" and as "technology". Thus the bioconservative critique of emerging biotechnologies is grounded in a view of technology as something that impinges on the human from an outside; this is implied in the very notion of technological "intervention". While in transhumanist discourse the human is presented as having a transcendent position vis-à-vis its environment, and as using technology to master that outside.

1.2   Beyond the Humanist Dualist Paradigm

A number of critical theoretical developments in the later part of the twentieth century have contributed to complicating this dualist paradigm and the rather simplistic view of human-technology relations that emerges from it. New perspectives in science and technology studies, media studies, anthropology, feminist studies and the philosophy of technology, have argued for richer conceptualizations of technology and technologies: as a political and cultural phenomenon (Feenberg 1991;

Haraway 1991; Winner 1980), as a social activity (Bijker et al. 1987; Callon and Latour 1992; MacKenzie and Wajcman 1985) and as mediating entities (Ihde 1993; Latour 1992, 1994), rather than as the human's "other". In the views of these theorists, the humanist dualist paradigm cannot account for the deep intimacy, the intricate enmeshing between humans and technology that has always been an integral part of human experience and that has become increasingly evident with the advent of many new technologies. In this sense the very proliferation of human-technology hybrid entities that biotechnologies are giving rise to, from the more iconic images of "designer babies", genetically modified corn and transgenic mice, to the less obvious (but no less hybrid) ones of cosmetically and cognitively enhanced humans, surrogate mothers and recipients of brain implants, are all evidence that this dualist paradigm can no longer be upheld. Indeed, the ontological divide that is drawn between human beings and technology becomes an obstacle to understanding the many ways in which humans and technology, but also subjects and objects, nature and culture, are interwoven today, and obscures the ways our interactions with technologies shape "what it means to be human" on a number of levels.

The starting point of this book is the recognition that insofar as the dominant discourses on the impact of emerging biotechnologies for what it means to be human are informed by precisely the humanist division that these technologies constantly undermine, they cannot provide an adequate conceptual basis from which to begin to address this question, and the many others that emanate from it. What is needed, then, is an alternative framework to begin with, one that acknowledges the heterogeneous, perhaps even emergent nature of human subjectivity, the active, mediating nature of technologies and the intricate enmeshing of both as an integral part of human experience. This book aims to develop such a non-humanist perspective.

1.3   Mapping the Posthuman

In recent years, the discussion on the philosophical and ethical implications of emerging biotechnologies and their significance for what it means to be human has converged around the evocative terms "posthuman" and "posthumanism". These terms will provide the theoretical landscape of this study. There has been a clear increase in scholarly interest in the posthuman over the last couple of decades. A number of readers and introductory guides focusing on the posthuman (alternatively incarnated as the cyborg), have been published, including but not limited to Judith Halberstam and Ira Livingston's *Posthuman Bodies* (1995), Chris Hables Gray's *The Cyborg Handbook* (1995), Neil Badmington's *Posthumanism* (2000), Fiona Hovenden, Linda Janes, Gill Kirkup, and Kathryn Woodward's *The Gendered Cyborg* (2000), Elaine Graham's *Representations of the Post/Human* (2002), Julian Savulescu and Bostrom's *Human Enhancement* (2009), Bert Gordijn and Ruth Chadwick's *Medical Enhancement and Posthumanity* (2008) and Cary Wolfe's *What is Posthumanism?* (2009). But an attempt to bring together the various types of posthumanist discourse that converse (or do not converse) across the field in a systematic and inclusive manner is still lacking.[1]

A mapping of the theoretical terrain via a characterization and classification of different types of posthumanist approaches, and further, an attempt to make them converse with one another, would be a helpful first step in bringing some clarity into the entangled relationship between emerging biotechnologies and notions of human nature. Indeed, even a quick look at the various definitions of the posthuman or posthumanism reveals that these terms mean very different things for different theorists. Nick Bostrom (2003), the transhumanist philosopher, defines the posthuman as someone who has basic capacities that greatly surpass those of humans in their present form. For the sociologist and philosopher of science and technology Andrew Pickering (2005), the posthuman refers to a new unit of analytical inquiry, that has emerged out of the coupling of the human and the non-human. For others, the post-human refers to a much less tangible or explicit entity. Francis Fukuyama's (2002) posthuman evokes a crisis, in which human nature and the social values that are based in it are under siege. And for others still, such as the sociologist Nicholas Gane (2006), the posthuman designates the opening up of a new critical culture, characterized by new forms of creative evolution that undermine the alleged purity of human nature. As might be expected with any terms in which "human nature" – its reconceptualization, fixing or delineation – is at stake, "posthuman" and "post-humanism" are highly contested.

An exploration of these terms, which will be the focus of Chap. 2, reveals four different types of posthumanist discourse that are exemplified by the definitions given above: a "dystopic", a "liberal", a "radical" and a "methodological" posthu-manism. Dystopic posthumanism is characterized by an objection to the use of technology to modify or enhance humans beyond broadly accepted natural and cultural limits. This includes what is often termed bioconservative literature (Annas 2005; Fukuyama 2002; Kass 1997; Sandel 2007) as well as critical defenses of humanism in the context of emerging biotechnologies (Habermas 2003). Liberal posthumanism is characterized by an endorsement of emerging biotechnologies for their perceived ability to allow humans to transcend their biological limits and enhance themselves at will. This includes the work of transhumanist theorists (Bostrom 2005; Hughes 2004; Kurzweil 2005; Moravec 1990; Savulescu 2007), and is common in other liberal approaches to new biotechnologies (Agar 2004; Buchanan 2011a, b; Harris 2007). As mentioned above, dystopic and liberal post-humanism represent the dominant approaches in the public debate on emerging biotechnologies.

---

[1] Some exceptions are Cary Wolfe's (2010) very good albeit brief mapping, developed as an introduction to the "Posthumanities" book series (see http://www.carywolfe.com/post_about. html); James Hughes' (2002) comprehensive "The Politics of Transhumanism" that can be found online at http://www.changesurfer.com/Acad/TranshumPolitics.htm; and Dale Carrico's (2006) "Technoprogressivism: Beyond Technophilia and Technophobia", a post that can be found on the "Institute for Ethics & Emerging Technologies" site, http://ieet.org/index.php/IEET/more/carrico20060812/.

Radical posthumanism is characterized by the view that emerging biotechnologies are contributing to a deconstruction of foundational discourses based in terms like "nature" and "the human". This interdisciplinary approach is informed by cultural theory, cyborgology, feminist studies and Science and Technology Studies (STS) (Badmington 2000; Balsamo 1996; Braidotti 2006; Graham 2002; Gray 1995; Haraway 1991, 1997; Hayles 1999; Stone 1995; Zylinska 2002). These theorists often view the idea of the co-evolution of humans and technology as liberating – not from the human species' historical bondage to nature and finitude in the sense that liberal posthumanists do – but from the notion that "human" and "nature" are fixed categories, ones that have been historically defined in opposition to their constitutive others. In many respects, radical posthumanism can be seen as a continuation of poststructuralist and early postmodern theory, that extends the postmodern critique of modernity and the Enlightenment – namely the anti-humanist critique of the unified, rational subject and the critique of dialectic logic – into an age of ubiquitous technoscience. It thus views the posthuman as providing a means of *political* resistance against the metanarratives of modernity and as having the potential to usher in a postmodern and post-anthropocentric era. This approach will be called radical because it calls for a radical rethinking of human ontology in light of emerging biotechnologies.

Finally, methodological posthumanism is characterized by an attempt to conceptualize analytical frameworks that can better account for the networks and zones of intersection between the human and the non-human. This includes STS scholarship (Latour 1992, 1999; Pickering 2005) and the newer generation of philosophers of technology (Ihde 1993, 2009; Verbeek 2005, 2011). Methodological posthumanism offers various frameworks for thinking the co-constitutive character of human-technology interactions, from "ontological relationality" (Ihde 1993, 2009), to "actor-network theory" (Callon and Law 1997; Latour 1992), "symmetry" (Latour 1993, 1999) and "manglings" (Pickering 1995). It also introduces two crucial notions for the analysis of posthuman technologies: an emphasis on *materiality*, or the study of the concrete development and formation of particular technologies and their impact on human experience (as opposed to more traditional transcendental perspectives of technology), and what is known as *technological mediation*, the understanding that technologies are not neutral instruments or intermediaries but rather active mediators that contribute to shaping the relation between users and their environment. While there are many philosophical implications involved here, methodological posthumanism can be seen more as an attempt to develop better conceptual tools for studying science and technology in society rather than developing a new posthuman ontology – hence the use of the term methodological for this approach.

This mapping of posthumanist discourse complements and aims to be more inclusive than some of the taxonomies that have been developed recently, namely by bringing together Anglo-American and "Continental" strands of thought on the posthuman. But several provisos need to be made here. First, as is the case with any such undertaking, the map is never the territory. Typologies necessarily simplify significant nuances and emphases that differentiate theorists, and they can never be exhaustive. The aim here is to offer working categories that can provide some clarity in a very multifaceted and complex field, and I believe these four types of posthumanism represent the main positions in the discussion. Second, none of the theorists mentioned here self-ascribe to the categories proposed. Other than the transhumanists and several methodological posthumanists, few of the theorists here would agree to having their approaches designated as posthumanist at all. This is true especially for dystopic posthumanists, who can in a sense be seen as "anti-posthumanists". But it seems clear that their negative appraisal of the posthuman takes part in posthumanist discourse nonetheless. This is also true of the cyborg theorist Donna Haraway, who has expressed her exasperation with the "posthuman", as a term that has been co-opted by transhumanists.[2] But it is precisely the need to differentiate between these understandings of the posthuman and to prevent its appropriation by any one group that justifies such a mapping. This means that there will not only be some inevitable simplifications and generalizations, but also that the attempt to establish these groupings as working categories will take place as they are being applied. In this sense this is a performative work. Thirdly, "humanism" itself is not an unambiguous concept. As many theorists have observed, it is often reduced to only one of its many versions (Halliwell and Mousley 2003; Soper 1986), and in its rich form, is already engaged with elements of anti-, counter- or posthumanism (Badmington 2000; Hardt and Negri 2000). But for all of its "bagginess", the greater part of posthumanist discourse converses with *one* account of humanism, which upholds the subject as a free, autonomous, self-contained being with clear boundaries that is detached from the empirical world. It is this account of humanism that dystopic posthumanism is fearful of losing, that liberal posthumanism attempts to extend, and that both radical and methodological posthumanism seek to overcome.

1.4 Non-Humanist Posthumanisms

The most obvious means of positioning these various types of posthumanism would be along an axis of celebration versus condemnation: do they embrace or object to the widespread use of emerging biotechnologies in light of the fact that they have significant implications for what it means to be human? But as shall be argued throughout this book, this is not the most meaningful or helpful axis of differentiation for moving forward in this debate. Rather, it is the humanist or non-humanist underpinnings of these approaches, regardless of their praise or skepticism, that should be emphasized, and that will be the most productive organizing theme for understanding how these approaches differ and relate to one another, and what they

---

[2] This also explains her more recent turn towards "companion species", see Gane (2006). The idea of this co-optation is also Hayles' (1999) argument in *How We Became Posthuman*.

have to contribute to thinking about implications of emerging biotechnologies for what it means to be human.

For both methodological and radical posthumanism the experience of being human is always shaped by our interactions with technology, and the reality we live in consists of a complex web of relations between the human, the world and the technologies that mediate between them, a network of human and nonhuman entities that is constantly in the making, constantly creating new realities based on the novel connections and associations being made. This is to say that both approaches, though they differ significantly in their philosophical background and understandings of the practical implications of their analyses, are based in a rejection of the humanist categorical distinction between autonomous human beings and a world of objects, which is seen as only one specific configuration of the relations between humans and the world. The humanist notion of an autonomous, fixed and unitary subject comes to light in this context as a by-product, an illusory effect of this division, rather than some true essence of human beings. Both these approaches also develop models of technology that reject the essentialism that underlies conventional models of technology – be they instrumental (as in liberal posthumanism) or substantive (as in dystopic posthumanism) – according to which either humans have a mastery over technology or technology has a mastery over humans. As such, they offer much-needed non-humanist alternatives to dystopic and liberal posthumanism. The exploration of these two approaches will occupy a large part of this study.

In addition, these approaches reject the overall pessimistic, romantic and transcendentalist view of technology as a dehumanizing and alienating force that characterizes classical philosophy of technology, so that they also offer an important contribution and revision of the critical approaches that precede them. Instead, and this proceeds from their anti-essentialism, they develop models of technology that allow for positive appraisals of technology, in which technology offers a form of engagement with the world. Radical posthumanism argues for a reflexive model of technology in which technologies are both seen as the product of human creativity and a force that shapes human existence. For radical posthumanists this implies a celebration of the political potential inherent in new technologies to overcome some of the most detrimental effects of modernity. For methodological posthumanism, the key notion of mediation replaces alienation as the central concept for analyzing technology, and this implies the need to conceptualize the *ambivalent* status of technology, which, though it may lead to a loss of involvement of humans in their environment in some instances, can also amplify and create novel forms of engagement. An underlying assumption of this study is that critiques based in the so-called dehumanizing effect of new biotechnologies are of little help for shedding light on the profound repercussions these technologies have, and that such pessimist presumptions prevent us from identifying many of the positive and enriching effects of these technologies – not just in the realm of medical treatment, but also for what it means to be human. This shift towards more positive understandings of contemporary biotechnologies, then, informs my search for a non-humanist model of human-technology relations.

1.5   The Shortcomings of Methodological and Radical Posthumanism

If radical and methodological posthumanism do offer better theoretical frameworks than dystopic and liberal posthumanism, however, they are not without significant shortcomings. In the framework of methodological posthumanism, these become apparent in any discussion on posthuman subjectivity. For methodological posthumanists, the prevalence of human/non-human couplings and networks indicates that humans do not necessarily have a monopoly on agency, intentionality or morality, which can be extended to artifacts, as something that is "delegated" to them, or inherently theirs (Latour 1992, 1999). Yet while it is clear that the freestanding intentional humanist subject cannot remain intact in this framework, a new, coherent model of what post-subjective subjectivity might entail is never clearly articulated by methodological posthumanists. Ultimately, it seems that breathing life *into* objects, so to speak, is more important for methodological posthumanists than delving into the implications of having breathed life *out of* subjects.

To be fair, it is somewhat unjustified to expect a coherent model of subjectivity from these theorists insofar as it is the need to develop new conceptual tools – they are *methodological* posthumanists – for the analysis of science and technology in society, rather than the deep philosophical implications this has for human subjectivity, that preoccupies them. In this sense this lacuna is not so serious. This is less the case, however, with radical posthumanism, of which I take up a critical examination on two levels. Regarding subjectivity, like methodological posthumanism, radical posthumanism shares the critique of humanism's dualist metaphysics and contributes to posthumanist discourse two aspects of poststructuralist theory: the ethical significance that is implied in the construction of new kinds of subjectivities and the political valorization that emerges from the dissemination of the autonomous, unitary subject.

But the significance for radical posthumanism of subjectivity as a platform from which to resist power gives rise to an incoherence: on the one hand, most radical posthumanists ascribe to the idea, expressed in Haraway's (1991) notion of the "informatics of domination", that the current formation of power is a post-disciplinary configuration that thrives on the collapse of binary thought, difference and multiplicity. On the other hand, the political potential that is identified in post-human figures of resistance such as the cyborg is said to lie precisely in the transgression of the system of binary oppositions that underlies Western patriarchal power apparatuses – in the ability to break down those boundaries that it is claimed have in any case already collapsed. It is not clear, in other words, how the multiple and fragmented nature of posthuman subjectivity, which can understandably act as a site of resistance to *modern* disciplinary power, can also embody the ideal form of resistance in a post-disciplinary or *postmodern* configuration of power that is itself multiple and fragmented. Another way of stating this ambivalence is that the mobile, fragmented, posthuman subject is simultaneously presented as a *symptom* of the contemporary configuration of power and as an *agent of resistance* to it.

In this context it is necessary to question what qualitative kind of impact the notions of hybridity, fragmentation and fluidity, so frequently celebrated by radical posthumanists, really have.

To be sure, this is not an uncommon concern among radical posthumanist theorists themselves, who are often preoccupied by the adoption by private biotech companies and public scientific discourse of a vocabulary of heterogeneity, flexibility and boundary transgression. But this inconsistency runs deep, through what we shall see is one of the underlying claims of radical posthumanism: the contention that emerging biotechnologies threaten to destabilize the modernist project by undermining foundational, essentialist categories such as the human and the natural. Put simply, the claim made by radical posthumanists is often that this potential is ultimately stifled, that if any "de-naturalization" of naturalized terms does take place, these are just as soon "re-naturalized" and "re-essentialized". I will explain in much greater detail how this complex process is understood. But what becomes clear is that radical posthumanism falls back onto a rather strictly dichotomous framework in this context – not for or against the use of these technologies as framed in dystopic and liberal posthumanism – but between the deconstructive or "post-modern" potential they embody and the disciplinary or "modern" uses they are put to. Drawing on some concrete examples of uses of emerging biotechnologies, I will argue that, while radical posthumanism is a very useful framework for shedding light on the shuffling around of foundational categories in our technological culture, it does not do enough to show how the so-called modern and postmodern co-exist in the context of emerging biotechnologies, giving rise to *new* understandings of notions like nature, the human and subjectivity in ways that undermine such a modern versus postmodern or disciplinary versus deconstructive framework.

1.6   Towards a Mediated Posthumanism

The close examination of methodological and radical posthumanism reveals that, although they offer important non-humanist alternatives to dystopic and liberal posthumanism, these approaches fail to capture significant aspects of the implications of emerging biotechnologies for notions like subjectivity, nature and human nature. Following this examination, a new perspective, "mediated posthumanism", will be developed that builds on the non-humanist basis that radical and methodological posthumanism establish, but aims to overcome their limitations while bringing together their valuable insights. This involves two main elements.

First, mediated posthumanism draws on the deconstructive readings of radical posthumanism, incorporating the idea that new biotechnologies have a tremendous destabilizing effect on taken for granted boundaries between the natural and the technological, thus undermining the classical humanist framework. But instead of framing these in a dialectic of deconstructive potential versus disciplinary or unifying praxis, this perspective sheds light on how these tendencies seem to coincide and intertwine on many occasions, engendering unexpected narratives of nature and humanness, of de- and re-naturalizations. Here I draw on a number of ethnographic works carried out by sociologists of health that illustrate how users integrate and normalize the use of new biotechnologies. As these studies indicate, ideas of genuine or authentic selfhood, ideas of biology as deterministic and of nature as fixed essence, intermingle and overlap in surprising ways with what seem to be conflicting ideas of a contingent or shifting self, of biology as open to transformation and of nature as technologically produced. A mediated posthumanist approach aims to account for this novel flexibility and richness, or duality of new meanings. Furthermore, in relation to radical posthumanism, a shift takes place here in terms of what the significant focus of research is. The important question that we need to ask becomes, not if and how terms like "human" and "nature" are re-naturalized or re-essentialized, but if the re-naturalizations that are taking place in specific cases are *positive* ones. I suggest that positive re-naturalizations may be characterized by their relocalization and coordination in creative and productive ways within new narratives of nature, identity and selfhood that take on at least momentary and context-specific intelligibility insofar as they hold together and can be functional for users. This shift of emphasis unquestionably opens up the discussion in new directions.

Secondly, mediated posthumanism extends the notion of "technological mediation" developed by methodological posthumanism – the idea that technologies are not mere modest means to an end but active mediators that help shape the relationship between humans and their world – into the realm of *bio*-technology. This gives greater depth to the notion that technologies are bearers of morality, insofar as the decisions taken in the framework of emerging biotechnologies are frequently moral ones – from those concerning the medication of what were once seen as personality traits to those concerning the lives of unborn fetuses. Furthermore, a mediated posthumanist perspective acknowledges that if technologies "interfere" with who we are, than this requires a rethinking of the status of subjects as well as of objects. This means continuing where methodological posthumanism "leaves off", by carrying through the transformative implications the notion of technological mediation has for subjectivity. I suggest that a fruitful direction in which to pursue this is Foucault's later work on "care of the self" (Foucault 1997, 2005). Here emergent reproductive, genetic and neuro-technologies can be interpreted as "technologies of the self", practices that take one's body, thoughts and conduct as a site for work, in the aim of transforming oneself into a specific moral individual. This use of Foucault is already quite widespread in the sociology of biomedicine (see especially the works of Nikolas Rose (2007)); but in this literature it does not engage with the idea of technological mediation. My suggestion is that, brought together, these two frameworks can yield important insights in the biotechnological context.

In this later work of Foucault's, ethics involves the ability to reflect on the processes by which we are endlessly constituted as subjects and requires that we develop an active relationship to the mediations that help shape the self. As Peter-Paul Verbeek has argued (2008, 2011), Foucault's line of thought is a very constructive place from which to begin thinking about how the technologically mediated character of life in our highly technological culture constitutes subjects in specific ways. By bringing together the notions of technological mediation and technologies of the self, mediated posthumanism provides an understanding of human-technology relations in which technology is neither a neutral tool nor a force that alienates humanity from itself, as in the liberal and dystopic posthumanist approaches, but something that is already part of the experience of being human. Technology is understood here as transformative without being deterministic, and the human relinquishes its traditional monopoly on agency yet retains a conscious, ethical relationship to its technological mediations. Subjectivity in this framework is an emergent property, that arises from interactions between various natural and technological, human and non-human fields, that is constantly being shaped and transformed by its engagements with biotechnologies. In the last chapter, this mediated posthumanist framework is used to explore the emergence of "genetically responsible selfhood" as a new mode of subjectivity that shapes how individuals think about themselves and that informs ethical decision-making that extends beyond the immediate medical realm.

1.7   Structure of the Book

The book is divided into eights chapters. Chapter 2, following the introduction, presents a review of the numerous and diverse definitions of the terms posthuman, posthumanism and humanism, and a mapping of posthumanist discourse along several axes – condemnation/celebration, historical/philosophical and humanist/non-humanist. The four different types of posthumanism that will be used throughout the book emerge from this mapping.

Chapter 3 reviews the state of the human enhancement debate and lays bare the many different arguments advanced by dystopic and liberal posthumanism concerning issues such as cognitive enhancement, "designer babies", genetic engineering and "cosmetic psychopharmacology". While these arguments are usually articulated in terms of risk and access, it is argued that this has the effect of obscuring the fact that what is at stake, for both approaches, is human nature. And while these seem to be conflicting views of human nature, it is argued that they are two versions of the humanist worldview and its dualist paradigm, which draws a strict separation between humans and their technologies.

Chapter 4 takes a closer look at radical and methodological posthumanism as the main candidates for a non-humanist alternative to dystopic and liberal posthumanism via the models of human-technology relations that they develop. It focuses on the notions of "technological mediation" and "originary prostheticity" that allow these approaches to move beyond the essentialist models of technology advanced by liberal and dystopic posthumanism. Technology in these frameworks, as something that is always already part of the experience of being human, can neither be seen as a neutral tool, as in the liberal posthumanist model, nor a force that alienates humanity from itself, as in the dystopic posthumanist model. These anti-essentialist and materialist models of human-technology interaction also allow these approaches to argue for more positive conceptualizations of technology than classical philosophers of technology, without falling into an uncritical technophilic assessment of technology.

For radical posthumanism this becomes the basis for a certain celebration of the political potential inherent in new biotechnologies to collapse the binary oppositions that underlie modern structures of power. For methodological posthumanism, this means articulating the ambivalent status of technologies, which can amplify new forms of engagement alongside the loss of known forms of engagement.

Chapter 5 steps back from the technological realm and takes a look at how the humanist dualist paradigm is also being challenged in current biological research. The examples discussed here include molecular bioscience, which is generating the view of an "analogue" body made up of flexible and mobile elements of genetic information that can be transferred between bodies and between species; complexity theory in evolutionary biology, which is promoting a dynamical view of the organism as an open system that actively participates and interacts with its environment; and molecular phylogeny, which offers a view of species and organisms as the results of endosymbiotic fusions and genetic flux between domains of life. This is to say that it might be possible to speak of a shift in these disciplines, from a "molar" formulation of the body or organism, understood as a self-contained, unified organic whole that is distinct from its environment, to a "molecular" body or organism, understood as a fragmented assemblage made up of transferable and translatable parts that depends much more on interactions with its surroundings. This biological form of originary prostheticity complements its anthropological counterpart that is articulated in Chap. 4.

Chapter 6 explores the transformations that subjectivity undergoes in the radical posthumanist and methodological posthumanist approaches, through a discussion of a number of philosophical perspectives on the subject via Heidegger, Latour, Deleuze and Haraway. These approaches have in common the assumption that the autonomous, fixed and unitary subject of liberal humanism is a by-product of the rigid separation of subjects and objects undertaken by modern metaphysics, rather than some true essence of human beings. This understanding allows for other configurations that may take into account the many ways in which humans and their technologies are interwoven. Subjectivity is reformulated in these views as an emergent property that arises from interactions between various human and non-human fields. The discussion on posthuman subjectivity brings to light the several significant shortcomings of methodological and radical posthumanism that were summarized above. This critique then serves as a platform to introduce mediated posthumanism.

Chapter 7 pushes forward the critique of radical posthumanism by applying it to a reading of reproductive technologies and the notion of "technologized nature". Here Gilles Deleuze and Félix Guattari's (1977) "schizoanalysis" provides a useful means of framing radical posthumanist readings of emerging biotechnologies, where the "schizophrenic" tendency of biotechnologies undermines the fixity of the category nature (deterritorialization) and its "paranoid" tendency contains this potential and channels it back onto normalizing categories (reterritorialization). The chapter looks at what happens to the notion of nature in assisted reproduction, as it is constantly de- and re-naturalized. I argue that rather than an ultimate re-naturalization, narratives and alternative ontologies do not cancel each other out in these contexts but are layered onto one another in novel reterritorializations that users often draw upon in strategic ways. Here nature is both "given" and "given to control", i.e. it incorporates both paranoid and schizophrenic tendencies, a duality which technologies continuously help to reconfigure. It is this dual or flexible nature which the mediated posthumanist approach aims to capture. With this understanding, the interesting question becomes not if and how re-naturalizations (or reterritorializations) take place but if these are positive or negative ones.

In Chap. 8 the mediated posthumanist perspective is developed to its full extent in an analysis of new genetic technologies. In the first part of the chapter the argument for the flexibility and richness of new understandings of nature in the context of assisted reproduction, developed in Chap. 7, is extended to the categories of "life" in the context of genomics and "subjectivity" in the context of neuro-technologies. Here paranoid trends such as genetic determinism and narratives of authentic selfhood coexist and intermingle with schizophrenic trends such as genetic complexity and a novel ontology of flatness. This analysis offsets popular critiques of "geneticization" and "genetic essentialism". In a second part I argue that this duality in current understandings of subjectivity expresses a shift in the kind of persons we take ourselves to be. This is manifested in a new mode of "genetically responsible subjectivity" that informs ethical decision-making that extends beyond the immediate medical realm. The analysis here brings together the notion of technological mediation, framing biotechnologies as forms of engagement with the world that have a moral dimension, and Foucault's later work on care of the self, so that these technologies can be understood as ethical practices that are deployed by individuals upon themselves in order to transform themselves in desired ways.

The structure of the book can thus be seen as moving from a general and predominantly theoretical discussion, on posthumanist discourse (Chaps. 2 and 3), followed by the role of technology (Chap. 4), the body (Chap. 5), and subjectivity (Chap. 6) in the context of posthumanism, to the application of the mediated posthumanist approach to more concrete analyses of specific neuro-, reproductive and genomic technologies. Each chapter questions the role that technology plays, in the various approaches, in defining the boundaries of the human, the subject and nature. While the identification of genetic responsibility as a new mode of selfhood marks somewhat of an end point to the book, I would like to think of it as a starting point, since it is only one example of the intriguing and multifaceted novel modes of human experience that are emerging in the current biotechnological landscape, and one that needs to be explored further in order to lay bare all the detrimental and beneficial aspects it entails. In this sense this study is only a preliminary step in this direction, by seeking to develop the theoretical approach that does this best.

References

Agar, N. (2004). *Liberal eugenics: In defence of human enhancement.* Oxford: Blackwell.
Annas, G. (2005). *American bioethics: Crossing human rights and health boundaries.* Oxford: Oxford University Press.

Badmington, N. (Ed.). (2000). *Posthumanism*. New York: Palgrave.

Balsamo, A. (1996). *Technologies of the gendered body: Reading cyborg women*. Durham/London: Duke University Press.

Bijker, W. E., Hughes, T. P., & Pinch, T. (Eds.). (1987). *The social construction of technological systems: New directions in the sociology and history of technology*. Cambridge, MA: MIT Press.

Bostrom, N. (2002). Existential risks: Analyzing human extinction scenarios and related hazards. *Journal of Evolution and Technology, 9*(1). http://jetpress.org/volume8/symbionics.html. Accessed 20 August 2013.

Bostrom, N. (2003). The transhumanist FAQ, Version 2.1. http://www.transhumanism.org/resources/FAQv21.pdf. Accessed 6 June 2013.

Bostrom, N. (2005). In defense of posthuman dignity. *Bioethics, 19*(3), 202–214.

Braidotti, R. (2006). *Transpositions: On nomadic ethics*. Cambridge: Polity Press.

Buchanan, A. (2011a). *Better than human: The promise and perils of enhancing ourselves*. New York: Oxford University Press.

Buchanan, A. (2011b). *Beyond humanity? The ethics of biomedical enhancement*. Oxford: Oxford University Press.

Callon, M., & Latour, B. (1992). Don't throw the baby out with the Bath school! A reply to Collins and Yearley. In A. Pickering (Ed.), *Science as practice and culture* (pp. 343–368). Chicago: Chicago University Press.

Callon, M., & Law, J. (1997). After the individual in society: Lessons on collectivity from science, technology and society. *Canadian Journal of Sociology, 22*(2), 165–182.

Carrico, D. (2006). Technoprogressivism: Beyond technophilia and technophobia. http://ieet.org/index.php/IEET/more/carrico20060812/. Accessed 14 Jun 2013.

Deleuze, G., & Guattari F. (1977). *Anti-Oedipus: Capitalism and schizophrenia* (trans: Seem, M., Lane, H.R., & Hurley, R). New York: Viking Press. Original edition, 1972.

Feenberg, A. (1991). *Critical theory of technology*. New York: Oxford University Press.

Foucault, M. (1997). Technologies of the self. In P. Rabinow (Ed.), *Ethics, subjectivity and truth: of the essential works of Michel Foucault 1954–1984 Vol. 1* (pp. 223–251). New York: The New Press.

Foucault, M. (2005). *The hermeneutics of the subject: Lectures at the Collège de France, 1981–1982* (trans: Burchell, G.). New York: Palgrave Macmillan.

Fukuyama, F. (2002). *Our posthuman future: Consequences of the biotechnology revolution*. New York: Farrar, Straus and Giroux.

Gane, N. (2006). "When we have never been human, what is to be done" interview with Donna Haraway. *Theory, Culture and Society, 23*(7–8), 135–158.

Gordijn, B., & Chadwick, R. (Eds.). (2008). *Medical enhancement and posthumanity*. New York: Springer.

Graham, E. L. (2002). *Representations of the post/human: Monsters, aliens and others in popular culture*. New Brunswick: Rutgers University Press.

Gray, C. H. (Ed.). (1995). *The Cyborg handbook*. New York: Routledge.

Habermas, J. (2003). *The future of human nature*. Cambridge: Polity.

Halberstam, J., & Livingstone, I. (Eds.). (1995). *Posthuman bodies*. Bloomington: Indiana University Press.

Halliwell, M., & Mousley, A. (2003). *Critical humanisms: Humanist/anti-humanist dialogues*. Edinburgh: Edinburgh University Press.

Haraway, D. (1991). A Cyborg manifesto: Science, technology, and socialist-feminism in the late twentieth century. In D. Haraway (Ed.), *Simians, Cyborgs and women: The reinvention of nature* (pp. 149–181). New York: Routledge.

Haraway, D. (1997). *Modest_Witness@Second_Millenium. FemaleMan©_Meets_Oncomouse™: Feminism and Technoscience*. New York: Routledge.

Hardt, M., & Negri, A. (2000). *Empire*. Cambridge, MA: Harvard University Press.

Harris, J. (2007). *Enhancing evolution: the ethical case for making better people*. Princeton: Princeton University.

Hayles, N. K. (1999). *How we became posthuman: Virtual bodies in cybernetics, literature and informatics*. Chicago: University of Chicago Press.

Hovenden, F., Janes, L., Kirkup, G., & Woodward, K. (Eds.). (2000). *The gendered cyborg: A reader*. New York: Routledge.

Hughes, J. J. (2002). The politics of transhumanism. http://www.changesurfer.com/Acad/TranshumPolitics.htm. Accessed 6 June 2013.

Hughes, J. J. (2004). *Citizen cyborg: Why democratic societies must respond to the redesigned human of the future*. Boulder/Colorado: Westview Press.

Hughes, J. J. (2009). Social pressures for technological mood management. *Free Inquiry, 29*(5), 28–32.

Ihde, D. (1993). *Postphenomenology: Essays in the postmodern context*. Evanston: Norhtwestern University Press.

Ihde, D. (2009). *Postphenomenology and technoscience*. Albany: State University of New York Press.

Kass, L. (1997). The wisdom of repugnance. *The New Republic, 216*(22), 17–26.

Kurzweil, R. (2005). *The singularity is near: When humans transcend biology*. New York: Viking.

Latour, B. (1992). Where are the missing masses? Sociology of a few mundane artifacts. In W. E. Bijker & J. Law (Eds.), *Shaping technology/building society: Studies in sociotechnological change* (pp. 225–259). Cambridge, MA: MIT Press.

Latour, B. (1993). *We have never been modern*. Cambridge, MA: Harvard University Press.

Latour, B. (1994). On technical mediation: Philosophy, sociology, genealogy. *Common Knowledge, 3*, 29–64.

Latour, B. (1999). *Pandora's hope: Essays on the reality of science studies*. Cambridge, MA: Harvard University Press.

MacKenzie, D., & Wajcman, J. (Eds.). (1985). *The social shaping of technology*. Milton Keynes: Open University Press.

McKibben, B. (2003). *Enough: Staying human in an engineered age*. New York: Times Books.

Moravec, H. (1990). *Mind children: The future of robot and human intelligence*. Cambridge, MA: Harvard University Press.

Pickering, A. (1995). *The mangle of practice: Time, agency and science*. Chicago: University of Chicago Press.

Pickering, A. (2005). Asian eels and global warming: A posthuman perspective on society and the environment. *Ethics & The Environment, 10*(2), 29–43.

Rose, N. (2007). *The politics of life itself: Biomedicine, power, and subjectivity in the twenty-first century*. Princeton: Princeton University Press.

Sandel, M. (2007). *The case against perfection: Ethics in the age of genetic engineering*. Cambridge, MA: Harvard University Press.

Savulescu, J. (2001). Procreative beneficience: Why we should select the best children. *Bioethics, 15*(5), 413–426.

Savulescu, J. (2007). In defence of procreative beneficence. *Journal of Medical Ethics, 33*(5), 284–288.

Savulescu, J., & Bostrom, N. (Eds.). (2009). *Human enhancement*. Oxford: Oxford University Press.

Soper, K. (1986). *Humanism and anti-humanism*. London: Hutchinson.

Stone, A. R. (1995). *The war of desire and technology at the close of the mechanical age*. Cambridge, MA: MIT Press.

Verbeek, P.-P. (2005). *What things do: Philosophical reflections on technology, agency and design*. University Park: Penn State University Press.

Verbeek, P.-P. (2008). Obstetric ultrasound and the technological mediation of morality: a postphenomenological analysis. *Human Studies, 31*(1), 11–26.

Verbeek, P.-P. (2011). *Moralizing technology: Understanding and designing the morality of things*. Chicago: Chicago University Press.

Winner, L. (1980). Do artifacts have politics? *Daedalus, 109*, 121–136.

Wolfe, C. (2009). *What is posthumanism?* Minneapolis: University of Minnesota Press.

Wolfe, C. (2010). Posthumanities. http://www.carywolfe.com/post_about.html. Accessed 7 Mar 2011.

Zylinska, J. (Ed.). (2002). *The cyborg experiments: The extensions of the body in the media age*. London/New York: Continuum.

Chapter 2
A Cartography of the Posthuman

Abstract This introductory chapter offers a comprehensive mapping of posthumanist discourse along three axes of differentiation: an optimistic/pessimistic axis, a historical-materialist/philosophical-ontological axis, and a humanist/non-humanist axis. It is argued that this last axis of differentiation, where humanism refers to a radical separation between human subjects and technological objects, is the most consequential one. Using these axes, four broad types of posthumanism are identified: "dystopic posthumanism", "liberal posthumanism", "radical posthumanism", and "methodological posthumanism".

Dystopic posthumanism is characterized by an objection to the use of technology to modify or enhance humans beyond broadly accepted natural and cultural limits. Liberal posthumanism is characterized by an endorsement of bio- and enhancement technologies for self-modification and self-improvement, grounded mainly in an individual rights framework. Radical posthumanism is characterized by the view that bio- and enhancement technologies, by undermining the fixity of categories like "nature" and "the human", contribute to a deconstruction of humanist and Enlightenment narratives based in human uniqueness and call for a radical rethinking of what it means to be human. Finally, methodological posthumanism is characterized by the development of analytical tools and frameworks that can (better) describe and highlight the zones of intersection and interaction between humans and technologies that play an essential part in human experience. These four approaches will become working categories for the rest of the book, and will be built upon in order to develop a final "mediated posthumanist" approach.

Keywords Posthuman • Posthumanism • Emerging biotechnologies • Transhumanism • Bioconservatism

The terms "posthuman" and "posthumanism" have become increasingly widespread over the past several decades in both academic and popular circles. Trendy, provocative, apocalyptic or celebratory, the posthuman seems to have become a catchphrase

T. Sharon, *Human Nature in an Age of Biotechnology: The Case for Mediated Posthumanism*, Philosophy of Engineering and Technology 14,
DOI 10.1007/978-94-007-7554-1_2, © Springer Science+Business Media Dordrecht 2014

for many things cyber, hi-tech and avant-garde and for any sign that we might be at the dawn of a new epoch, one in which we are losing or have already lost some essential tie to nature and have become open to technological modification. The common thread among the diverse and wide-ranging uses of these terms is the idea that advanced and emerging biotechnologies, from genomics to assisted reproduction to neuroscience, have an impact upon our very understanding of what it means to be human. Beyond this, however, it is a lack of consensus as to what the posthuman refers to, and what its implications for "what it means to be human" are, that characterizes posthuman discourse.

A quick glance at the posthuman repertoire attests to this, beginning with the diverse forms of punctuation used to connect or disconnect between the "post" and the "human". Though most commonly today "posthumanism" and "posthuman" will be written as one word, one also finds both terms conjoined by hyphens or backslashes ("post-human", "post/human") or written as two entirely separate words ("post human"), and written with or without one or more capital letters ("post-Human", "Post-Human"). This may not be inconsequential. Neil Badmington (2004), locating his work within a tradition fascinated by the significant impact small punctuation marks can have on meaning, has argued that the presence or lack of punctuation marks conjoining or separating the "post" and the "human" can be the sign of entirely different understandings or treatments of the posthuman.[1] At the least, it is the sign of an underlying vagueness that surrounds the term: an *inherent* vagueness, due to the very elusiveness of the terms "humanism" and "human" that the prefix "post-" cannot relate to as something fixed; as well as an *extrinsic* vagueness, since even when there is agreement regarding the meaning of "humanism" and "human", as we shall see, the posthuman can have quite dissimilar and even conflicting undertones and implications.

A closer look at some definitions of the posthuman, which can refer to a tangible, physical entity, a historical condition, or a new critical perspective, makes this even clearer. Nick Bostrom, one of the leading transhumanist thinkers today, defines the posthuman as "someone whose basic capacities so radically exceed those of present humans as to be no longer unambiguously human by our current standards" (2003). And on the Transhumanist FAQ website, we find that posthumans,

> will be persons of unprecedented physical, intellectual, and psychological ability, self-programming and self-defining, potentially immortal, unlimited individuals. Posthumans have [*sic*] overcome the biological, neurological, and psychological constraints evolved into humans.[2]

Robert Pepperell (1995), focusing more on the nature of human consciousness, has written about the posthuman as a general convergence of organisms and

---

[1] Badmington's essay is built around the significance of the backslash used by Elaine Graham in her work on the "post/human", inspired by Derrida's obsession with the quotation marks found throughout Heidegger's texts in *Of Spirit: Heidegger and the Question* (1989), and Jean-François Lyotard and Eberhard Gruber's analysis of the hyphen in *The Hyphen: Between Judaism and Christianity* (1999). He claims that by introducing the oblique, Graham succeeds in bringing to the fore more challenging and subtle meanings of an otherwise relatively simple term.

[2] See http://humanityplus.org/learn/philosophy/faq#answer_20.

technology that renders them indistinguishable, as well as the end of a period of social development known as Humanism. What he calls the "Post-Human condition" refers to the effect of the collective impact technologies have on what constitutes a human being and "our sense of human existence". In terms not so dissimilar to the transhumanists', the social and political philosopher Francis Fukuyama, author of *Our Posthuman Future* (2002a), relates the posthuman to the possibility of modifying, of enhancing, "our complex evolved natures", by means of biotechnology. Though, for Fukuyama such a disruption in the continuity or unity of human nature is perceived as a looming crisis, since the concepts of justice, morality and rights that are grounded in our understanding of human nature will also be undermined.

Fukuyama's malaise indicates that it is often the disruption of traditional conceptual frameworks that is engendered by the posthuman as a variation on the human as a biological organism which is central in the use of the term. Thus, a number of theorists view the posthuman as the opening up of a new conceptual and critical space. In "What's Wrong with Posthumanism", for example, Herbrechter and Callus define posthumanism as a discourse, which

> articulates our hopes, fears, thoughts, and reflections at a post-millenarian time haunted by the prospects of technology's apparently essential and causal link with the finiteness of the human as biological, cognitive, informational, and autonomous integrality" (2003: paragraph [g]).

Andrew Pickering (2005) uses the term posthuman as a new unit or object of analytical inquiry that surfaces when we pay attention to the constant coupling or tuning of human subjects and non-human objects, and that challenges traditional units of analytical inquiry in the sciences ("things") on the one hand and in the humanities and social sciences ("people") on the other. For Catherine Waldby the posthuman refers to,

> an effect of the slippage involved in effacing or naturalizing [the human's technogenetic network of production] in the interests of maintaining the amour propre of the human, moments of disjuncture which leave this technogenic network exposed and available for critical analysis. (2000: 43)

And Elaine Graham understands what she calls the "post/human" as an "interrogative marker, a critical cue", that "both confounds but also holds up to scrutiny the terms on which the quintessentially human will be conceived" (2002: 11 and 36).

The plurality of definitions presented in this short overview indicates that the posthuman, as an entity, as a historical condition or as a conceptual space, evokes uncertainty and contestation, where much is at stake. About the future, since embedded in conceptualizations of the posthuman are usually claims to what normative and desirable humanity should be, and what kinds of values and rights it would be grounded in. But also about the past (and the present), since conceptualizing the posthuman inevitably presupposes, challenges or calls to defend understandings of notions like subjectivity, nature and authenticity that lie at the core of our philosophical, predominantly humanist, tradition. This chapter aims to map the complex and contested terrain of posthumanist discourse by typifying various styles or models of posthumanism along three axes of differentiation that cut across it (see Fig. 2.1). These are (1) a pessimist/optimist axis, (2) a historical-materialist/

AXIS 1

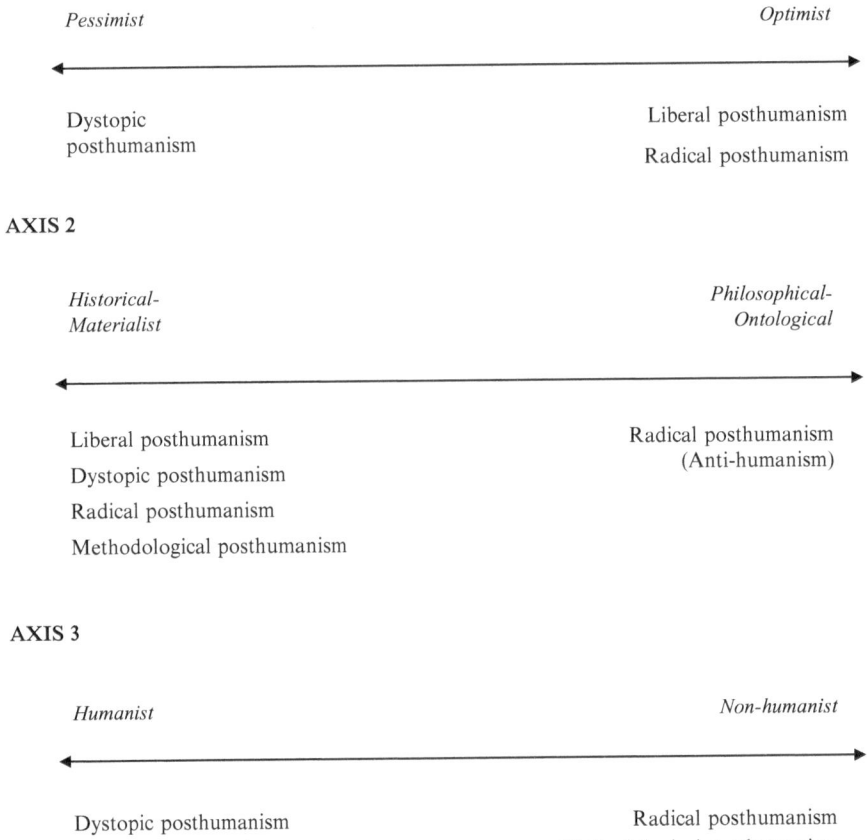

AXIS 2

AXIS 3

Fig. 2.1 Axes of differentiation and types of posthumanism

philosophical-ontological axis, and (3) a humanist/non-humanist axis.[3] Using these axes, four broad discourses of the posthuman are identified: "dystopic posthumanism", "liberal posthumanism", "radical posthumanism", and "methodological posthumanism". Dystopic posthumanism is characterized by an objection to the use of technology to modify or enhance humans beyond broadly accepted natural and cultural limits. Liberal posthumanism is characterized by an endorsement of bio- and enhancement technologies for self-modification and self-improvement, grounded mainly in an individual rights framework. Radical posthumanism is characterized by the view that bio- and enhancement technologies, by undermining the fixity of categories like "nature" and "the human", contribute to a deconstruction of

---

[3] This mapping has been in part inspired by Cary Wolfe's (2010) introduction to the Minnesota Press "Posthumanities" series, where he offers a short mapping based on pessimist/optimist, historical/ontological and "dry"/"wet" categorizations.

humanist and Enlightenment narratives based in human uniqueness and call for a radical rethinking of what it means to be human. Finally, methodological posthumanism is characterized by the development of analytical tools and frameworks that can (better) describe and highlight the zones of intersection and interaction between humans and technologies that play an essential part in human experience. The aim of this introductory chapter is not to map all of the existing literature on biotechnologies and posthumanism, but to identify broad themes and types of posthumanist discourse, and to offer ways of comparing and contrasting them. These four approaches will become working categories for the rest of the book, which I will build upon in order to develop an additional "mediated posthumanist" approach.

2.1 The Pessimist/Optimist Axis

One of the easiest ways to make sense of different approaches to emerging bio- and enhancement technologies is to place them along an axis of pessimism/optimism. In this framework, approaches range from explicit rejections of the desirability of human enhancement and a call to ban or restrict enhancement technologies to solely therapeutic practices, to varying degrees of tolerance of and regulation of their development and use, to arguments for the overriding utility and desirability of enhancement and the view that we have a moral duty to provide and use these technologies.

2.1.1 The Pessimistic Pole: Dystopic Posthumanism

One of the most pronounced positions in the discussion on human enhancement and the posthuman is the view that the modification of human beings using biotechnology to overcome what are generally perceived as human biological and cultural limitations is dangerous and in some way immoral (Annas et al. 2002, 2005; Fukuyama 2002a, 2004; Habermas 2003; Kass 1985, 1997, 2002, 2003; McKibben 2003; Rifkin 1998; Sandel 2007; Smith 2004). "Dystopic posthumanism" as this approach shall be identified in this book, is characterized by the moral claim that human enhancement is intrinsically wrong, and the political claim that the state should be involved in banning or restricting it.[4] The central thesis of this approach is that enhancement and biotechnologies threaten human nature, and since a fixed, stable human nature is essential to our notion of human dignity and its legal counterpart, human rights, we must do everything possible to protect it. Fukuyama

---

[4] "Dystopic posthumanism" is also known as "bioconservatism", as many advocates of human enhancement designate it. In the United States, dystopic posthumanist theorists come from two groups that are usually in disagreement: religious conservatism (see for example the Center for Bioethics and Culture, http://www.thecbc.org/, and the Center for Bioethics and Human Dignity, http://www.cbhd.org/), and liberal environmentalism (see for example the Center for Genetics and Society, http://www.geneticsandsociety.org/, the Council for Responsible Genetics, http://www.gene-watch.org/index.html, and the ETC Group, http://www.etcgroup.org/en/).

(2002b) calls for the establishment of a new regulatory agency that will have "statutory authority over all research and development". George Annas et al. (2002) calls for the establishment of a "human species protection treaty". And Jürgen Habermas (2003) upholds each human's "right to a genetic inheritance immune from artificial intervention". This does not mean that dystopic posthumanists do not believe that emerging biotechnologies can have any potential benefits. Biotechnology is a "devil's bargain" in the words of Fukuyama (2002a: 8), where subtle harms are intertwined with obvious benefits, and where the state needs to introduce a regulatory legislative framework to separate would-be legitimate from illegitimate uses of biotechnology, usually along the problematic therapy vs. enhancement distinction. Indeed, in general dystopic posthumanists have little faith in the libertarian claim that the free market can or even should be allowed to decide what uses biotechnology will be put to. This is because what is at stake in the introduction of enhancement technologies is usually not so much notions of personal freedom and individual choice, but humanity and human nature as we know it.[5]

Leon Kass for example, one of the leading and most outspoken dystopic posthumanists, who chaired the President's Council of Bioethics under George Bush between 2001 and 2005, has dedicated much of his writing on new biotechnologies to their "dehumanizing" effects. In the beginning of *Life, Liberty and the Defense of Dignity* (2002), Kass writes,

> Human nature itself lies on the operating table, ready for alteration, for eugenic and psychic "enhancement", for wholesale re-design. In leading laboratories, academic and industrial, new creators are confidently amassing their powers and quietly honing their skills, while on the street their evangelists are zealously prophesying a posthuman future. (2002: 4)

For Kass, technologies like cloning, reproductive selection, regenerative medicine and life extension all represent a first step in the engineering of humans towards an "inhuman" future. This is mainly due to what is seen as a relentless desire for mastery over nature and human nature as the main impetus towards human enhancement. The posthuman here implies a final, technical conquest of man over his own nature. This is not necessarily in contradiction with the view of the posthuman endorsed by some liberal posthumanists, as we shall see. But for Kass, as for other dystopic posthumanists, this "voluntary dehumanization" (1985: 71) both makes the human less than what she was intended to be by nature – by transforming her into raw material – and more than what she was intended to be – by locating the meaning and source of life within her own will and power. He writes:

> Here … is the most pernicious result of our technological progress – more dehumanizing than any actual manipulation or technique, present or future: the erosion, perhaps the final erosion, of the idea of man as noble, dignified, precious, or godlike, and its replacement with a view of man, no less than of nature, as mere raw material for manipulation and homogenization. (2000: 86–87)

---

[5] For example, in the very popular report formulated by the President's Council of Bioethics entitled *Beyond Therapy: Biotechnology and the Pursuit of Happiness* (2003), in each instance of a move "beyond therapy" that was identified, special concerns were expressed regarding the deformation of humanity not by governmental use of biotechnologies, from above, but from below, from consumer endorsement of their use for individual desires.

This is to say that there is something in the state of finitude that mortality and the limits on physical malleability have imposed on human beings which is inherently good. These finite limits bring out the best in humans, the "noble" and the "dignified", since it is in coming to terms with the pain and suffering that they entail that humans derive meaning and virtue. And it is, further, in the creation of virtue that ensues from this acknowledgement of human limits and finitude that humans rise above the rest of nature.

Michael Sandel's opposition to human enhancement also draws on the dangers of the desire for mastery that is encouraged by and that underlies the use of biotechnologies for human modification and improvement. In his tellingly named book *The Case Against Perfection* (2007), Sandel argues that human enhancement technologies will diminish our appreciation of the "giftedness" of life – the understanding that human life is always partly beyond our control. In his own words:

> The deeper danger is that they (i.e., enhancement and genetic engineering) represent a kind of hyperagency, a Promethean aspiration to remake nature, including human nature, to serve our purposes and satisfy our desires. The problem is not the drift to mechanism but the drive to mastery. And what the drive to mastery misses, and may even destroy, is an appreciation of the gifted character of human powers and achievements. To acknowledge the giftedness of life is to recognize that our talents and powers are not wholly our own doing, nor even fully ours, despite the efforts we expend to develop and exercise them. It is also to recognize that not everything in the world is open to any use we may desire or devise. (2007: 26–27)

Alongside the loss of a sense of giftedness of our lives, Sandel argues that human enhancement will also lead to a loss of certain central human values. Humility, he explains, will be replaced by excessive pride as we become the sole source of our own achievements, just our sense of solidarity will erode as we become the only ones to blame for our failures.

In *Our Posthuman Future* (2002a), Fukuyama, who like Kass was also a member of the President's Council, also expresses apprehension that biotechnology poses a significant threat to human nature, and further to the rights and values, even the political order, that are grounded in it. Fukuyama maintains that we need to protect the "full range of our complex, evolved natures against attempts at self-modification" (2002a: 172) in order to preserve the basis of human dignity and rights. This is because for Fukuyama there is a direct link between human nature and human dignity and rights: solid political structures and societies are not created *ex nihilo*, but are derived from innate behavioral characteristics, ambitions and drives of the individuals that make them up. Thus a natural moral sense, he argues, has evolved over time as demonstrated in a range of emotive responses that is "species-typical" (2002a: 140–143). Any alteration in our shared biological heritage, such as in our genetic endowment, or in our species-specific cognitive mechanisms, will cause a rupture in our commonly shared human nature and the world it has helped create. One of Fukuyama's main concerns, for example, is the gap that will inevitably appear between technologically enhanced posthumans and unenhanced humans. In a society that will be divided into the "GenRich", the strata of those who will be able to afford purchasing "good" genes to improve the genetic make-up of their offspring, and the "GenPoor", who will certainly lose the status of moral dignity they currently possess (2002a: 154), the principal of equal moral dignity will be a thing of the past. In this posthuman future, any basis for a political, legal or moral appeal to a notion of our shared humanity will collapse.

Jürgen Habermas, the German critical philosopher, has also recently taken up the case against enhancement, focusing on genetic interventions, embryo research and preimplantation genetic diagnosis (PGD). In his attempt to provide solid grounds for rejecting the use of such technologies for enhancement purposes, he also appeals to human nature. Habermas argues that the foundation and justification for morality and procedural justice, in which society is configured in such a way as to respect the autonomy of the individual and her values and conception of the good life, lies in a prior, ethical self-understanding of the species, a minimal self-understanding of ourselves as human. This "species ethics" includes three important elements: the understanding that we are the undivided, autonomous authors of our own lives; that we approach others as the individual authors of *their* lives, i.e. a recognition of equality; and that we seek to live with others who acknowledge us as self-individuating autonomous beings, in a society that protects and nourishes our right to be so. This clear understanding of what makes us human is necessary in order for us to respect ourselves and others.

Biotechnologies, especially genetic intervention at the reproductive level, are a threat to this species ethics because they involve an undermining of one's capacity to see one's self as the undivided and responsible author of one's own life. Enhancement can thus be harmful both to individuals and society. Habermas writes:

> We cannot rule out that knowledge of one's own hereditary features as programmed may prove to restrict the choice of an individual's life, and to undermine the essentially symmetrical relations between free and equal human beings. (2003: 23)

An individual who has been genetically enhanced at the will of an other, cautions Habermas, and who is aware of this fact, will feel in some way diminished and less authentic, having been tailored towards someone else's expectations. Her life will be seen as an artifact, and her existence instrumentalized to fulfill the desires of another human being. Consequently, she would to some extent be incapable of feeling responsible for her life and of locating herself as part of the moral community of humans. As for Fukuyama, intervention on this basic biological level of the human species has profound consequences for humanity, morality and the foundations of our liberal democracy.

2.1.2 The Optimistic Pole: Liberal Posthumanism

At the other end of this spectrum, optimistic accounts of the posthuman abound. These accounts inform both what I will call "liberal posthumanism" and "radical posthumanism". But while liberal and radical posthumanism can be seen as analogous to the extent that they both welcome to varying degrees the advent of the posthuman, as we shall see the promises that the posthuman holds for these two approaches are completely different.

The approach that will be identified as liberal posthumanism in this book is comprised of transhumanist scholarship (Bostrom 2005, 2009; Hughes 2004; Kurzweil 2005; Moravec 1990; Savulescu 2001, 2005, 2007, 2010) as well as other liberal and rights-based arguments for the permissibility of human enhancement (Agar 2004; Buchanan 2011a, b; Glover 2006; Harris 2007). For liberal posthumanists, the right to use enhancement technologies is an extension of individual freedom, choice and self-determination – upheld as inalienable aspects of what it means to be human. In this sense, individuals are entitled, as human beings, to alter any aspects of their biology they choose to as long as this does not directly harm others. Restricting genetic or neurological modification is thus interpreted as an assault on essential freedoms that are at the core of the liberal democratic ideal.[6]

Transhumanism is an international movement that advocates the feasibility and desirability of improving the human condition by using new technologies to transcend the biological limits of the human species.[7] Philosopher Nick Bostrom, the co-founder of the World Transhumanist Association (WTA, recently renamed "Humanity+") and one of its most outspoken members, defines transhumanism as:

> (1) The intellectual and cultural movement that affirms the possibility and desirability of fundamentally improving the human condition through applied reason, especially by developing and making widely available technologies to eliminate aging and to greatly enhance human intellectual, physical, and psychological capacities. (2) The study of the ramifications, promises, and potential dangers of technologies that will enable us to overcome fundamental human limitations, and the related study of the ethical matters involved in developing and using such technologies. (Bostrom 2003)

This popular, highly technocratic form of posthumanism sees our current era of rapid technological advance as the transitional phase between our human past and a posthuman future.[8] New technologies it is claimed, including genetic engineering, neural-computer integration, regenerative medicine and cognitive enhancing pharmaceuticals, may allow us to overcome the biological, neurological and psychological constraints that have been evolved into humans, and are all seen as preludes to

---

[6] It is interesting how akin liberal and dystopic posthumanism are here. Interestingly, not only do both approaches believe their positions best reflect public opinion, but they both employ a strategy that allows them to claim legitimacy as the moral position that best defends current dominant political and social values. As Roache and Clarke (2009) suggest, transhumanists are so confident of their position as the real defenders of liberal democratic values that they believe that in the absence of a marked policy shift they will win the human enhancement debate by default.

[7] The term "transhuman" was originally coined by the evolutionary biologist Julian Huxley in 1957 in his work *New Bottle for Old Wine*, anticipating a historical condition in which human beings have obtained the capabilities to impact their own biological evolution through technological means, and to describe an ethics based around the mapping of and alteration of human physical, intellectual and aesthetic possibilities. Huxley writes,

> The human species can, if it wishes, transcend itself – not just sporadically, an individual here in one way, an individual there in another way, but in its entirety, as humanity. We need a name for this new belief. Perhaps transhumanism will serve: man remaining man, but transcending himself, by realizing new possibilities of and for his human nature (1957: 17).

[8] The "trans" of transhumanism alludes to the desire to transcend, but also to this *transitional* period.

an era when people will routinely enhance their brains, improve their bodies and perhaps live forever.

There are many versions of transhumanism and transhumanists come in a much wider variety than can be listed here.[9] But some common characteristics include: a *technophilic* attitude that locates technology as a solution to human societal and biological problems; a basic belief in *technogenesis*, the assumption that technology is involved in a spiraling dynamic of co-evolution with human development; and the claim that there is a *moral obligation* to pursue enhancement when the technological means to do so exist. Early roots of the transhumanist movement can be found in the organizations for life extension, cryonics, space colonization, science fiction, and futurism of the 1970s and 1980s.[10] In the last two decades the group has gathered force and coalesced around organizations like the Extropy Institute (now defunct), an especially optimistic branch of transhumanism created by Max More and Tom Morros in the early 1990s; the Alcor Life Extension Foundation that advocates cryonics, the Palo Alto-based Foresight Institute, founded by Eric Drexler in 1986 for research in molecular nanotechnology, the Singularity Institute for Artificial Intelligence, founded in 2000, the Future of Humanity Institute of the Faculty of Philosophy at Oxford University directed by Nick Bostrom, the Institute for Ethics and Emerging Technologies co-founded by Bostrom and James Hughes, and the WTA, or "Humanity +", founded in 1998 by Bostrom and David Pearce, which currently boasts over 6,000 members. The movement's general philosophy and views have been expressed in manifestos, declarations, "FAQs" online and in many international conferences.

While extreme versions of this discourse may remain on the margins of public culture, somewhat less extreme versions are often the stuff of popular media, in for example the *New York Times'* technology and health sections, the BBC.com's "future" page and popular films such as *The Matrix* or *Terminator*. What's more, if admittedly some transhumanist notions such as life-extension, intelligence augmentation and mind uploading may sound like techno-utopian fantasies, it is important to understand that transhumanist objectives are being pursued through mainstream science, and by researchers at some of the most distinguished research centers in the

---

[9] A distinction can be made between "libertarian" transhumanism, as expressed by the Extropian Institute or computer scientist Marvin Minsky and "democratic" or "technoprogressive" transhumanism, as expressed in the Institute for Ethics and Emerging Technologies, and individuals such as philosopher Pierre Lévy, Anders Sandberg, Nick Bostrom, AI developer Ben Goertzel, longevity biologist Aubrey de Grey, and singularitarian Eliezer Yudkowsky. The WTA incorporates both currents. Even milder forms of transhumanism can be found in the works of writers such as Gregory Stock (2002) Michio Kaku (1998) and Kevin Warwick (2002). But even this distinction is too general since it is on specific issues (technological risks, individual liberty, etc.) that transhumanists differ.

[10] Some "transhumanists avant la lettre" include British biochemist J.B.S. Haldane, who argued for the benefits that could ensue from controlling our genetics in his 1923 essay "Daedalus: Science and the Future." Haldane's essay inspired a number of such works by British scientists, such as J.D. Bernal's *The World, the Flesh and the Devil* (1929) and Olaf Stapledon's *Last and First Men* (1931).

world, including MIT, the Los Alamos National Laboratory in New Mexico and Oxford University.[11] For example, both Marvin Minsky, the co-founder of MIT's AI laboratory, and Hans Moravec, the founder of the world's largest robotics program at Carnegie Mellon University, have argued for the theoretical feasibility of mind uploading, wherein a person's mind and personality could be emulated by a computer. In his book *Mind Children: The Future of Robot and Human Intelligence* (1990) Moravec argues that the age of carbon-based life is coming to an end, and will be followed by the domination of intelligent machines. The only way for humans to survive the inevitable twilight of human dominance in this scenario is to become machines themselves, by uploading their consciousness into computers. Ray Kurzweil (1999, 2005), the proliferous inventor and futurist, argues in several books that our biological evolution is on the verge of being superseded by our technological evolution. "Technology," he writes, "is evolution by other means" (1999: 16). For Kurzweil, human evolution follows a simple pattern by which the power of technology expands at an exponential rate, and is currently on the verge of a radically accelerating era of change unlike anything seen before, which will culminate in the "Singularity".[12] In this near future, AI, advanced nanotechnology and cybernetics will have contributed to a new, post-biological era, where humans will merge with computers and exist as disembodied software or higher forms of intelligent life, making mortality, historically related to the longevity of our "hardware", a thing of the past.

Other philosophers and bioethicists who are not transhumanists can also be grouped under the label of liberal posthumanism. These are authors who either argue in defense of human enhancement based on the core liberal value of freedom of choice, or against any general ban on enhancement that would be grounded in appeals to human nature, as something that may preclude individual conceptions of what makes a good life. In his book *Enhancing Evolution: The Ethical Case for Making Better People* (2007), John Harris uses a utilitarian framework to argue that our moral assessment of possible enhancements should be based on a rejection of harm and a recognition of the benefits that may derive from them. Human enhancement, he determines, is a good thing when it makes lives better; for individuals, for social policy and for a less than optimal human genetic heritage ("if it wasn't good

---

[11] Membership of the transhumanist movement also includes a host of cyberpunks, self-help gurus, nanotech venture capitalists and New Age enthusiasts, but I will be referring solely to transhumanist academics and scientists (though the identification of such varied subcultures with transhumanism is a fascinating cultural phenomenon in itself). Likewise, I will not linger on the numerous literary roots of the transhumanist movement, nor of radical posthumanism. These are located namely in the science fiction and cyberpunk fiction genres, including authors like Philip K. Dick, William Gibson, Bruce Sterling, Damien Broderick, Arthur C. Clarke, Isaac Asimov and Vernor Vinge among many others.

[12] The term "singularity" in the transhumanist context was introduced by science fiction writer and mathematician Vernor Vinge in a 1993 essay "The Coming Technological Singularity", which predicted that computers would be so powerful by 2030 that a new form of superintelllligence would emerge. Vinge compared this future point in history to the singularity at the edge of a black hole: a boundary beyond which old rules no longer apply.

for you", he puts bluntly, "it wouldn't be called enhancement" (2007: 9)). And they are not only morally permissible, but in some cases are even morally obligatory. Jonathan Glover, in *Choosing Children: The Ethical Dilemmas of Genetic Intervention* (2006), also employs a version of Mill's harm principle alongside the notion of human flourishing, arguing that limitations on parental liberty in the form namely of a restriction of preimplantation genetic diagnosis (PGD), should be considered only to safeguard human flourishing. If parents have the possibility to remove an obstacle to their children's flourishing – such as blindness and deafness, which he identifies as such following some examination of what constitutes a disability and its impact on the capacity for human flourishing – than they owe it to their children to do so. Further, it is possible that in light of technological advances, this duty will pertain not only to curing "disabilities", but to the enhancement of capacities in healthy children.

Others may claim, like Nicholas Agar, in his earlier book in defense of reproductive freedom *Liberal Eugenics: In Defense of Human Enhancement* (2004), that we do not have a moral duty to develop and use enhancement technologies, but that we must, if we want to remain consistent with the moral values of liberal democracy, tolerate and permit their use. Respect for individual conceptions of what makes a good life and which characteristics are desirable in a child commit us to resist any form of "authoritarian eugenics", Agar argues, but to defend what he calls "libertarian eugenics", the "procreative visions" of individual parents.[13] Allen Buchanan, in two recent works on the question of posthumanism, *Better than Human: The Promise and Perils of Enhancing Ourselves* (2011a) and *Beyond Humanity? The Ethics of Biomedical Enhancement* (2011b), also positions himself not so much in defense of human enhancement as such, but against an all-out rejection of enhancement technologies, namely one that would be based in an appeal to an essential human nature or a perfect balance of nature, and that forecloses a consideration of the pros and cons of various individual cases of enhancement. Drawing on arguments that are shared by many liberal posthumanists, as shall be discussed in the following chapter, Buchanan contends both that human evolution is suboptimal and unintended, so that there is no reason, a priori, not to try to improve it if the means to do so exist, and that human enhancement technologies do not differ essentially from many other enhancement techniques that humans have been practicing for millennia, such as numeracy, literacy and science.

---

[13] Agar's most recent book, as its title suggests, *Humanity's End: Why We Should Reject Radical Enhancement* (2010), is much less tolerant of human enhancement. Here Agar focuses on "radical enhancement" or enhancement practices such as mind uploading and certain types of germline engineering, that do not just seek to improve human attributes and abilities, but to raise them to levels that greatly exceed what is currently possible. Agar objects to this type of enhancement, which, unlike moderate enhancement, he argues, has the potential to generate experiences that unenhanced humans will no longer be able to engage with psychologically, to produce a species of posthumans that would not share the values and relationships of unenhanced humans, making moral reciprocity impossible, and ultimately resulting in conflict between both that would most likely end in the elimination of the unenhanced.

2.1.3 The Optimistic Pole: Radical Posthumanism

The approach that will be identified in this book as "radical posthumanism" can also be located at the optimistic pole of this first axis, where the advent of the posthuman is seen in positive terms. Like liberal posthumanism, radical posthumanism sees the blurring of boundaries that emerging bio- and enhancement technologies involve, between the human and the non-human, the natural and the technological, as potentially liberating. But rather than a liberation from the human species' historical bondage to nature, finitude and death, this is a form of liberation from the very notion that "the human" is a fixed category. Radical posthumanism views the posthuman in positive terms not because it suggests a transcendence or a replacement of a suboptimal human with a better and improved one – but because it indicates that the category of the human has always been (though this is increasingly evident today) inherently unstable and has never coincided with the supposed terms of its naturalization.

In a sense, radical posthumanism shares with dystopic posthumanism the idea that the technoscientific developments of the past decades have disturbed how we think about "the human" and "the natural", that they pose a threat to what we consider to be a human "essence" and the many values that are grounded in it.

But for radical posthumanists this is not a process to be feared and prevented, since it is a sign of the long-anticipated collapse of the liberal humanist project, in which the claim for an ontological difference between humans and the rest was a discursive practice that functioned "to domesticate and hierarchize difference within the human (whether according to race, class, gender) and to absolutize difference between the human and the non-human" (Halberstam and Livingstone 1995: 10). In this approach, bio- and enhancement technologies are seen as opening up, so to speak, both the physical body of the human and the theoretical concept of human nature. For radical posthumanists, these technologies thus offer the occasion to rethink the human in ways that may be more realistic, as an entity that emerges from within complex interrelations with its environment, and in ways that may be more ethical, as a being that emerges from within significant interrelation with others – be they human or non-human.

Radical posthumanism is an interdisciplinary approach informed by poststructuralist and postmodern theory, science and technology studies (STS) (Haraway 1991a, b; 1997; Latour 1993; Waldby 2000), cultural studies (Badmington 2000; Graham 2002; Gray 1995; Hayles 1999; Lyotard 1991; Stone 1991), and feminist, gender and queer theory (Balsamo 1996; Braidotti 2006b; Halberstam and Livingstone 1995). These theorists emphasize the destabilizing effects emerging biotechnologies have on many of the foundational categories of Western thought, notions like the human, nature, subjectivity and authenticity. They view them as contributing to the collapse of distinctions between what have traditionally been ontologically separate domains, like nature and technology or organism and machine, and to a deconstruction of foundational discourses based in "nature" and "the human". As the *material* instantiations of what were recently mainly *conceptual* claims of critical theory, these technologies and the hybrid entities they are giving rise to are seen as complementing the political promise of the postmodern franchise with a valuable technological impetus.

In many respects, radical posthumanism can be seen as a continuation of poststructuralist and early postmodern theory, that extends the postmodern critique of modernity and the Enlightenment – namely the anti-humanist critique of the unified, rational subject and the critique of dialectic logic that functions by reducing plurality and multiplicity to binary oppositions – into an age of ubiquitous technoscience. But radical posthumanism should not be seen as entirely analogous to postmodern theory, since it also places a strong emphasis on materialism (material bodies, physiological processes and more precisely embodiment), as part of the larger trend towards more concrete analyses in contemporary philosophy of science, sociology of science and feminism. Thus, while many of the central problematics of postmodern discourse remain, radical posthumanism redirects the primary focus of theory from the forces of social or linguistic construction to the extra-discursive and biological elements of human experience. Furthermore, radical posthumanism can be seen as complementing earlier critical theory with a positive assessment of technology, which it views as offering the potential for breaking with modernity.

Donna Haraway's "A Cyborg Manifesto" (1991a) can be seen as the founding text of this radical type of posthumanism.[14] The manifesto was originally written as a platform for a new socialist-feminism that, unlike most feminist theories of the time, would not conceptualize gender or "Woman" in relation to holistic unities and universal, totalizing theories. If the production of totalizing theory fails to grasp most of reality at any given moment in history, argued Haraway, than this is particularly true of the present time – and this is where Haraway's manifesto extends beyond an engagement with 1980s feminism and transforms into a political program for the posthuman era. Haraway argues that in our current age of advanced technologies we are witnessing the breakdown of three crucial boundaries that have hitherto been taken for granted: between the animal and the human, between the organism and the machine, and between the physical and the non-physical. These breakdowns, she argues, have rendered the classical humanist framework, in which the human and the inhuman, the natural and the unnatural, are held in binary opposition, obsolete. While Haraway does not explicitly use the term "posthuman", it is along these key boundaries that the normative "human" was traditionally delimited and its designated others excluded, so that their collapse marks the current era as irredeemably posthuman.

The cyborg appears precisely where these boundaries and others associated with them are transgressed.[15] It is a blasphemous, ironic and rebellious figure that deconstructs the humanist myth of wholeness and organicism. Grounded in hybridity and contingency, it is a creature that incarnates and embraces ambiguity and difference,

---

[14] The manifesto's complete title is "A Cyborg Manifesto: Science and Technology and Social-Feminism in the Late Twentieth Century" and was originally published in *Socialist Review* 80: March/April 1985.

[15] The term cyborg, short for "cybernetic organism", was coined in 1960 by NASA researchers Manfred Clynes and Nathan Kline (1995). It was originally construed as an enhanced human being who could survive the effects of long-term exposure to the weightlessness and artificial environments of outer space. Their research focused on the cybernetic or symbiotic approach of astronaut and spacecraft as interpenetrated systems that shared energy and information.

and proposes a radical form of inclusive politics based on affinity rather than identity, that cuts across traditional categories of difference such as race, gender and class.

Following the "Cyborg Manifesto", Haraway has focused on ways in which technoscientific developments, namely molecular genetics, immunology, and ecoscience, produce unique hybrid entities that challenge divisions between nature and culture, human and non-human and active subjects and passive objects. With the development of transgenic organisms such as the Oncomouse for example, one of Haraway's favorites, the idea of genetic integrity/unity of the organism is called into question. These entities are taken as evidence that the process of characterizing nature as Other, a project of policing borders and boundaries, is no longer tenable (Haraway 1997), and thus embody a significant potential to destablilize our traditional understandings of nature and subjectivity. As Rosi Braidotti has written of Haraway's "techno-monsters", they,

> contain enthralling promises of possible re-embodiments and actualized differences. Multiple, heterogeneous, uncivilized, they show the way to multiple virtual possibilities. The cyborg, the monster, the animal – the classical "other than" the human – are thus emancipated from the category of pejorative difference and shown in an altogether more positive light. (2002: 243)

Since the appearance of Haraway's manifesto, an abundance of literature forming what can be seen as the corpus of radical posthumanism has focused on how novel technologies contribute to the development of new discourses of identity along the human-technology interface, thus revealing the inherently prosthetic nature of human identity (Balsamo 1996; Bukatman 1993; Dery 1994, 1996; Featherstone and Burrows 1995; Graham 2002; Gray 1995; Mitchell and Thurtle 2004; Wolmark 1991; Zylinska 2002, to name but a few comprehensive works). The human, in many of these works, is viewed as an ongoing process of technological and anthropological evolution, as evidenced by recent advances in digital and biotechnologies. And if it is made, the argument goes, than it can be made differently and more ethically: both in a fashion that incorporates all those who, because of their gender, race, or other attributes, have historically been excluded from the dominant definition of the "human", and in a fashion that takes into consideration the interactions between humans and their environments, artifacts and tools, between (human) self and (non-human) other, that mutually constitute identity formation and any notion of self. In this approach, then, the inherent potential of emerging bio- and enhancement technologies to redraw boundaries between humans and the rest provides a means of political resistance against the metanarratives of modernity, and hope for a more ethical and sustainable future.

2.2   The Historical-Materialist/ Philosophical-Ontological Axis

While both liberal and radical posthumanism view the advent of a posthuman era as a positive event, then, their understanding of what the subject of a posthuman era will – or should – be, greatly differ. I will return to this discussion in more detail shortly, but would like first to introduce another axis that cuts across the posthuman landscape, between a historical-materialist understanding of the posthuman and a philosophical-ontological understanding. Classifying accounts of posthumanism as either of these does much more violence to the richness of the narratives at hand than the previous axis. This is because both poles are intertwined in ways that the previous positive or negative assessments were not, and at any moment either pole can be seen as a theoretical condition for its opposite. Furthermore, historical-materialist accounts can always be seen as expressions of certain presumed philosophical assumptions, in which case this distinction would relate only to how explicit these presumptions are. Nevertheless, this classification is important because it identifies the posthuman as a periodizing device on the one hand, as a term that refers to some historically specific event, and, on the other, as an approach that critically responds to this period. Alongside this it also distinguishes between accounts of the posthuman that are more concerned with the presence of non-human or non-biological components alongside the human, and those accounts that are more concerned with emerging models of subjectivity or with a more philosophical engagement with the question of the human itself.

2.2.1 The Historical-Materialist Pole: Dystopic, Liberal and Radical Posthumanism

Historical-materialist understandings of the posthuman view it as emerging within a specific historical time and as characterized by specific technological innovations. Both dystopic and liberal posthumanism ascribe to a historical understanding of the posthuman, since the emerging technologies of the present (and anticipated technologies of the future) are a precondition of the posthuman. In some transhumanist narratives this historical process will even unfold according to calculable timelines (Kurzweil's prediction for the integration of human and machines stands at 2045). For liberal posthumanists, importantly, the posthuman is usually framed in evolutionary terms, identified with the emergence of a new phase in the evolution of the human species, that will be driven by technology and intentionality rather than natural, random adaptive processes. Gregory Stock, Bill Clinton's biotechnology advisor in the 1990s, begins his book *Redesigning Humans: Our Inevitable Genetic Future* (2002), with the statement,

> We know that Homo sapiens is not the final word in primate evolution, but few have yet grasped that we are on the cusp of profound biological change, poised to transcend our current form and character on a journey to destinations of new imagination (2002:1)

The notion of a coming Singularity, a popular theme among liberal transhumanist writers (Kurzweil 2005; Moravec 1990; Vinge 1993), also implies a turning point, where a historical merge between humans and robots or digital technologies will utterly transform what it means to be human.

Dystopic posthumanists articulate similar claims about being on the verge of a profound transformation, at a unique moment in history. They are often prompted by a sense of urgency, in which the time to act is now. At the same time, this is a materialist understanding of the posthuman because any ontological shift, a shift in the nature of the human – or our understanding of the human – is preceded by a biological mutation. Thus, even though dystopic renderings of posthumanism such as Fukuyama's and Kass' focus on a fundamental potential alteration of human nature, this alteration is the result of an encounter with specific contemporary biotechnologies. In this use of the term, the posthuman is the result of the impact of new digital and biotechnologies on the human body, germline and psyche, resulting in various degrees of machinic symbiosis. In this view, there is an increasing internalization, incorporation or assimilation of material technologies, in the forms of prostheses, implants and synthetic drugs, by the organic body.

Radical posthumanism also refers to the posthuman as a historical contingency, though such a categorization of radical accounts of posthumanism to one pole of the historical-materialist axis is problematic because, as we shall see shortly, radical posthumanism is informed by a deep philosophical-ontological posthumanism.[16] The idea here is that the *quantity* of human and non-human connections is a historical novelty, and that the occurrence of such relations has become so frequent that a posthuman vocabulary has necessarily developed. In the introduction to *The Cyborg Handbook*, one of the anthologies of cyborg theory, Chris Hables Gray asks:

> But haven't people always been cyborgs? At least back to the bicycle, eyeglasses and stone hammers? … The answer is, in a word, no. Certainly, we can look back from the present at some human-tool and human-machine relationships and say "Yes, that looks very cyborgian", but this is only possible because of hindsight. … Cyborgian elements of previous human-tool and human-machine relationships are only visible from our current point of view. In quantity, and quality, *the relationship is new*. (1995: 6, emphasis added)

For Haraway (1991a, 1997) the technological advances of the late twentieth century are a key catalyst to the breakdown of previously rigid boundaries between human and machine, nature and culture, reality and non-reality, that make our historical moment posthumanist. In *The War of Desire and Technology at the Close of the Mechanical Age* (1995), Allucquere Rosanne Stone argues that we are at in the midst of a paradigm shift from the mechanical age to the virtual age, in which technology comes to be viewed as natural. And for Katherine Hayles (1999), the posthuman refers to a historically specific construction that recently emerged from the changing constellation of media, technology and culture. She traces the emergence of the "informational posthuman", which she sees as the prevalent form of posthumanism today, back to a specific historical event, the Macy Conferences on Cybernetics held between 1943 and 1954, and to a specific scientific model,

---

[16] In fact, attributing this distinction to radical posthumanism is *highly* problematic, and touches on a problem that will analyzed in greater detail in Chap. 7, namely, that radical posthumanism maintains both that a posthuman ontology is specific to the present, in light of our increasing encounters with technology, *and* that it better reflects our ontology in a non-historically specific way.

post-war cybernetics and its principles of circular causality and disembodied information. Haraway's cyborg and Hayles' posthuman, are, furthermore, presented as anticipated and logical steps in our evolution.[17]

2.2.2   The Historical-Materialist Pole: Methodological Posthumanism

The fourth and final posthumanist approach identified in this mapping is what shall be called "methodological posthumanism", introduced only here because it cannot be meaningfully framed in terms of the pessimist/optimist axis. Methodological posthumanism is characterized by the attempt to develop analytical tools that can conceptualize the inter-relationality of humans and technologies and that can fore-ground this inter-relationality as a significant aspect of what it means to be human. This approach is informed by STS scholarship (Akrich 1992; Bijker et al. 1987; Callon 1986; Latour 1992, 1999; Pickering 1995, 2005) and contemporary philosophy of technology (Ihde 1990, 1993; Verbeek 2005, 2011; Winner 1980). The important claim made by both is that technologies are not merely functional instruments that help humans realize their intentions, but that they actively contribute to the shaping of those intentions, and the human values and capacities that determine them.

This understanding has prompted many of the theorists who can be grouped under the methodological posthumanist label to focus their attention on how specific technological artifacts shape and modify human behavior, human perception, human decision-making and human identity. Artifacts, for example, as many STS scholars have argued, can be seen as having a "script" written into them (Akrich 1992; Latour 1992), that prescribes users how to act when using them: irritating seat-belt warnings can enforce the law on buckling up, automatic doors determine the speed at which people will walk through them, and bulky hotel key rings direct hotel guests towards the front desk to be dropped off before they step out. Technological artifacts, it is argued, can also shape how humans experience the world; they are a means of experience that consitutes a relation to the world. As the philosopher of technology Don Ihde (1990, 1998) has illustrated in numerous works, relations between human beings and the world take place "via" technological artifacts: eyeglasses, heating systems, computer mice, and telescopes, are all arti-facts that enable humans to experience and perceive different aspects of reality; they *mediate* human experience of the world. For some philosophers of technology arti-facts can even induce moral change (Swierstra et al. 2009; Swierstra et al. 2010) and contribute to a shaping of our morality (Verbeek 2005, 2011). Peter-Paul Verbeek suggests that by actively contributing to the coming about of moral actions and moral consideration surrounding those actions, technological artifacts can be seen

---

[17] The works of the media theorists Friedrich Kittler (1999) and the sociologist Niklas Luhmann (1995) can also be identified with this emphasis on the historical particularity of the phenomenon of posthumanism and a focus on particular technological developments, as Wolfe (2010) argues.

as "morally charged". The obstetric ultrasound, for example, is never merely a passive object used as an instrument to look into the womb. It opens up specific interpretations (of the fetus in terms of antenatal diagnostics) and it generates specific situations of moral choice (of terminating or carrying to term a pregnancy).

The influence that technological artifacts can have on human conduct and experience, these theorists argue, signifies that they are more than the inanimate objects they are usually conceived to be. In their materiality, they actively contribute to shaping and modifying human behavior in ways that challenge their depiction as lifeless things. This activity of technological artifacts indicates that technologies can to some degree be seen as possessing agency. This provocative proposition will be analyzed in more detail in Chap. 4. But what it means in terms of posthumanism and a methodological posthumanist approach is that humans and technologies interact in ways that are profoundly intertwined and seamless, and that it makes more sense to look at what happens at the zones of intersection between the human and the non-human than to obstinately pry apart their relative share in the behavior, identity and moral decision-making that ensues from them. Methodological posthumanists thus offer various frameworks for conceptualizing those zones of intersection, including "ontological relationality" (Ihde 1993, 2009), "actor-network theory" (Callon and Law 1997; Latour 1992), "(generalized) symmetry" (Callon 1986; Latour 1993, 1999) and "manglings" (Pickering 1995). In these frameworks humans and technologies, alongside other actors like institutions, social groups, devices and laboratories, come together to act as a whole, and the unit of analysis shifts from either things or people to a hybrid, posthuman entity that compromises both. For Bruno Latour (1999) for example, when a human shoots someone with a gun, it is the hybrid third entity, composed of a fusion of human and artifact, that is responsible for the action, rather than either of the individual "actants" that make up this network. Similarly, in his study on the decline of the scallop population in St. Brieuc Bay in north-western France, Michel Callon (1986) shows how fishermen, scientists and scallops alike, as equal actors, form networks and associations in order to translate their will and shape their world.

Insofar as human beings are not the only ones that "act" in such networks, but artifacts, devices, animals, and *matter* in general, do too, methodological posthumanism views materiality as a crucial aspect of human and social activity. For this reason methodological posthumanism can be positioned on the historical-materialist pole of our second axis. But the centrality of materiality in this approach does not signify so much that posthuman, hybrid collectives are a historical specificity of our age, since this has always been the case. Rather, it is because matter "matters" (Law 2010) in such a way that much of the analysis carried out by methodological posthumanists focuses on concrete technologies in specific contexts. For STS scholars, the emphasis on the diversity of technologies and the attempt to analyze the institutional, cultural and social contexts of human-technology relations in the form of individual case studies has been from the outset a methodological axiom. In the philosophy of technology, this focus on specific, concrete technologies is a more recent development, often referred to as an "empirical turn" (Achterhuis 2001). Here theorists have turned their attention away from the study of technology as a monolithic and rather abstract phenomenon, of technology with a "big T", that was characteristic of classical philosophy of technology and that often inspired particularly dismal depictions of modern technology (Ellul 1965; Heidegger 1977a; Jonas 1979). Instead, contemporary philosophy of technology subscribes to a pluralistic vision of technology, of a variety of technologies that in their concrete materiality each influence human experience in different ways, and each of which deserves separate analysis.

At the heart of the methodological posthumanist approach is thus a sensibility to the relationality of humans and non-humans and to materiality as two central features of the experience of being human. While there are many philosophical and ontological implications involved here, methodological posthumanism can be seen more as an attempt to develop better analytical tools for studying science and technology in society than an attempt to develop a new posthuman ontology. To a certain extent it can be seen as a descriptive more than a prescriptive or normative approach. And this is where it significantly differs from radical posthumanism, although both approaches, as we shall see shortly, share very similar assumptions.

2.2.3 The Philosophical/Ontological Pole: The Anti-Humanist Roots of Radical Posthumanism

For radical posthumanism, as we have seen, the notion of the posthuman implies a recognition that "what it means to be human" is always inextricably bound up with the environment, technological or other, in which "being human" takes place, and that this entanglement in turn implies a co-evolution of humans and technology. If for radical posthumanism this co-evolution may be more palpable in today's biotechnological, cybernetic and digital age, so that this approach does have a historicist aspect to it, this co-evolution is not exactly seen as a historical particularity. The human's entanglement with its (technological) environment has always been the case for radical posthumanism. But this does not only call for a rethinking of the analytical tools with which we describe and study human-technology relations; it has profound philosophical implications and points to the need for the formulation a new *post*human ontology. The theoretical roots of radical posthumanism in the "anti-humanism" of French critical theory are key here.

The anti-humanist tradition is characterized by the view that all humanism is ideological, and that the notions of "man" and "human nature" in the abstract are convenient fictions.[18] In its structuralist and poststructuralist revival anti-humanism

---

[18] The origins of this tradition can be traced back to the works of Marx, Nietzsche and Freud, who all argued that the human, as a coherent, unified subject is in one way or another the "effect" of causes located outside of herself. In Marx's materialist account of history, subjectivity (like social life) is seen as the effect of one's material conditions of existence. For Nietzsche, the subject is an effect of a vast array of competing instincts, desires, drives, beliefs, and capacities; it is a multiplicity rather than a pre-given unity. And in Freud's psychoanalytic theory, the human is largely driven by unconscious, irrational drives and desires.

placed itself in opposition to any thought that centered on the human subject as the primary analytical category, that posited a pregiven, unified subject or an unchanging human essence that precedes social operations, and contended that individuals are constituted as subjects by social practice grounded in discourse (Althusser 1992; Deleuze and Guattari 1977; Derrida 1976; Foucault 1989; Lacan 1977; Lévi-Strauss 1963). Anti-humanism aimed at dethroning, or at least at radically decentering the subject which had dominated the philosophical tradition stemming from Descartes through Sartre as merely an effect of language, culture, desire or the unconscious. Like Nietzsche, who saw the "death of God" as a unique opportunity for humans to surpass themselves, the demise of this humanist subject is understood by anti-humanists as a liberation from the idea of the human as a fixed essence and a transcendental ideal. Indeed, this will be taken up as the political potential inherent in radical posthumanism.

Foucault, for example, in *The Order of Things* (1989), argues that the humanist figure of "man" is the product of a specific historical episteme that emerged at the end of the eighteenth century with the rise of a new order of modernity. This new figure, both subject and object of the human sciences, is shown to be a novel discursive invention, rather than an eternal, natural occurring phenomenon. Foucault shows how fixed conceptions of human nature contribute to the rise of scientific normalization and social discrimination as powerful discourses of modernity and exposes humanism as the epistemological basis of disciplinary society. Thus, if the "arrangements of knowledge" that produced the idea of this human subject were to disappear – and Foucault contends that in light of the development of "counter-sciences" (psychoanalysis, ethnology and linguistics) they are – so would "man". The often-cited wager with which Foucault concludes *The Order of Things* is indeed a lyrical preface opening on to new configurations of the (post)human:

> As the archaeology of our thought easily shows, man is an invention of recent date. And one perhaps nearing its end. … If those arrangements were to disappear as they appeared … then one can certainly wager that man would be erased, like a face drawn in sand at the edge of the sea. (1989: 387)

For Deleuze and Guattari (1977), the humanist subject, as an autonomous and unified in-dividual is also seen as an illusion that is fostered and supported by systems of repression, normalizing discourses and institutions. But while Foucault undertakes a decentering of the humanist subject through a critical archaeology and genealogy that reduces the subject to an effect of discourse and disciplinary practices, Deleuze and Guattari introduce a "schizoanalytic" destruction of the ego and the superego in favor of a dynamic unconscious. As its title implies, in *Anti-Oedipus* (1977) their schizoanalytic brand of anti-humanism attacks the conforming and desire-repressing subject-production of psychoanalysis and seeks to liberate the prepersonal realm of desire, the libidinal flows that run below the conditions of identity. With Deleuze and Guattari, an emphasis on the co-extensivity of the body with its environment or territory, the enmeshment of the human in its larger biological and ecological context, indicates that the classical humanist division of the world along a human vs. non-human hierarchy cannot explain the rich couplings of human and non-human, the "machinic assemblages" that inform the individual as multiplicity. As we shall see in Chap. 5, this Spinozist-based ethology and the alternative, non-anthropocentric models of subjectivity put forward by Deleuze and Guattari (1987) associates them, more than any other poststructuralist theorists, with the contemporary theorists of radical posthumanism.

Expanding on early anti-humanists, the liberal humanist subject has, of course, been deconstructed by a number of approaches that also help form the theoretical milieu of radical posthumanism. Feminist theorists have pointed out that the liberal humanist subject has been constructed as a white European male that takes on a universality that suppresses women's voices (Irigaray 1985; Bordo 1993; Butler 1999). And postcolonial theorists have taken issue with the very idea of a unified, consistent identity (Spivak 1988; Bhaba 1994). In light of the importance of these *theoretical* antecedents of radical posthumanism, it becomes clear that the actual material encounters between humans and technology are somewhat less significant for radical posthumanism than the new models of subjectivity that are emerging from this encounter; models of cyborg affinity, models of machinic being, models of prosthetic hybridity and of posthumanity as inextricably bound up in relationality.

2.3 The Humanist/Non-Humanist Axis

The philosophical orientation that informs radical posthumanism leads us to our final, and what will be most consequential, axis of differentiation, and to the perhaps belated question of what the prefix "post-" in posthumanism refers to. Manifestly, this question is far from being straightforward. One reason for this seems to be the facile interchangeability of the terms post*human* and post*humanism*, where "human" supposedly relates to a biological entity, and where "humanism" supposedly relates to a theoretical and historically specific discourse. The assumption here being that if we were to avoid interchanging theses terms, we might achieve some clarity by grounding these terms in the more familiar distinction between a physical or biological realm and a theoretical or discursive one. The question becomes even more problematic, however, when we take into consideration that the biological/discursive distinction is itself not clear-cut. This is not only to say that bodily experiences can be the effect of particular cultures and historical periods, or even that our biology might rest on a metaphysics, or even that our biology can determine other levels of human experience. As feminist theory has commonly argued, our biological understanding of what the human is helps shape how we, as humans, relate to the humans, non-humans and world around us, all the while our ethical or philosophical understanding of our relations with our environment and the experiences they base help form the figure of the human we delineate in biological terms (see for example Grosz (1994) and Fausto-Sterling (2000)).

The interchangeability, then, between the terms human and humanism, where humanism is understood as a way of thinking about the human, and as an epochal discourse derived from a specific kind of thinking about the human (again, there can be no real precedence here), is not so much the sign of an intellectual laziness or confusion, but an indication, once more, of the complex nature of the very task of defining posthuman/ism. To be more precise, the question that should be posed is not just "what does the prefix 'post-' in posthumanism refers to?", but both "what kind of theoretical discourse about the human informs the biological understanding of the 'human' that the 'post-' in 'posthuman' refers to?", and "what kind of biological understanding of the 'human' informs the theoretical discourse about the human that the 'post-' in 'posthumanism' refer to?" – along with the stipulation that both questions must be posed together. In order to be reasonably consistent, a posthumanist account would have to respond to both questions with similar answers.[19]

2.3.1 A Multitude of Humanisms?

One should expect to find as many types of humanism as can be found posthumanisms. As Kate Soper (1986) and Tony Davies (1997) have cautioned in their well-known introductions to humanism, humanism can be a very elusive concept. According to Soper, the meaning of the term varies depending on its cultural context, namely, if we are "speaking English" or if we are "speaking French". Thus in the Anglo-American tradition, humanism is identified with the promulgation of secularism, while in the continental tradition it is identified with that notion of an essential, universal humanity that the anti-humanist movement challenges. Soper also enumerates three types of humanism with regards to the "anthropocentrism" commonly attributed to humanism. An "instrumentalist" humanism that assumes an opposition between an "external" reality and conscious human subjects, leading to an instrumentalist attitude towards nature which exists for and in the service of human ends. An "idealist" humanism, by which the world exists only insofar as it is reflected upon and understood in thought, i.e. by virtue of "Man's" conceptualization of it. And a "dialectical" humanism, that views humanity's relationship to nature as a totality, in which the world is transformed by humanity and humanity is shaped by its existence in the world.

Hardt and Negri (2000), in their influential work *Empire*, also speak of more than one humanism. They deconstruct traditional readings of modernity and offer instead the idea of a "modernity in crisis", a centuries' long struggle between two modes of modernity, each accompanied by very distinct traditions of humanism. They identify the first mode as a radical revolutionary process that is characterized by what they call the discovery of the "plane of immanence", that gave rise to a series of

---

[19] One could quote John Locke here:

> to be a man, or of the species of man, and have the essence of a man is the same thing. Now, since nothing can be a man, or have a right to the name man, but what has conformity to the abstract idea the name man stands for; nor any thing be a man, or have a right to the species, but what has the essence of that species; it follows that the abstract idea for which the name stands, and the essence of the species, is one and the same. (1823: paragraph 12)

philosophical, scientific and political developments stretching from the thirteenth to the sixteenth centuries that put an end to the medieval conception of being. These developments include an affirmation of humanity's power in the world, the shifting of knowledge from the transcendent to the immanent plane, a reappropriation by humanity of the "powers of creation" that medieval transcendence had consigned exclusively to the heavens, and a refoundation of authority on the basis of a human universal. Representative of this mode of modernity are the secularizing project of Renaissance humanism and its attack on transcendence, and Spinoza's philosophy of immanence which put humanity and nature in the position of God just as it posited that the laws of human nature were the same as the laws of nature as a whole.

This mode of modernity and its discovery of the plane of immanence was countered, Hardt and Negri explain, by a reactive attempt to re-establish ideologies of command and authority by redirecting the new image of humanity to a transcendent plane of order. This was achieved via three types or mechanisms of mediation, they suggest, that are articulated in the works of Descartes, Kant and Hegel: human knowledge becomes achievable only through the reflection of the intellect, nature and experience become recognizable only through the filter of phenomena, and the ethical world becomes communicable only through the schematism of reason (2000: 78–79). In the humanist project of this mode of modernity, there is a reinstatement of the conception of the human as separate and above nature, insofar as, through these mechanisms of mediation, the transcendence of God is transferred to the human.[20] European modernity, in Hard and Negri's scheme is defined by the crisis born of the struggle between these two modes and its temporary and incomplete resolution in the formation of the modern state, which succeeds in transcending and mediating the plane of immanent forces and imposing rule and order. Hence, what is most commonly identified as "modernity" and "humanism", they maintain, is really only the second, because triumphant, counterrevolutionary mode of modernity.

In *Critical Humanisms* (2003), Martin Halliwell and Andy Mousley also contest a monolithic view of humanism whereby one version of humanism is taken to represent the whole. They suggest an alternative taxonomy of humanistic thought based on a *pluralism* of humanisms, that includes no less than eight different versions: a "romantic", "existential", "dialogic", "civic", "spiritual", "pagan", "pragmatic" and "technological" humanism. Halliwell and Mousley argue that until the emergence of the anti-humanism of French critical theory, humanism indicated something much more diverse, amorphous, complex or "baggy", than it does at present. It was the anti-humanist movement's attempt to challenge humanist ideology, they claim, that introduced an identification of what were perceived as underlying premises that could designate humanism as a unified, cohesive philosophy, of which classical or liberal humanism became the representative form. "We are not simply suggesting that

---

[20] As Nietzsche, and Foucault after him observed:

is it not the last man who announces that he has killed God, thus situating his language, his thought, his laughter in the space of that already dead God, yet positing himself also as he who has killed God and whose existence includes the freedom and the decision of that murder? (Foucault 1989: 385) critical theory's version of humanism is a pure invention", they explain, "but that humanism has been tidied up, packaged and streamlined by some anti-humanists in such a way as to negate its actual diversity" (2003: 3).

To make matters more complicated, in addition to their project of re-evaluating a "critical" account of humanisms which would take into consideration both the diverse tradition of humanism and the post-foundationalist critique of anti-humanism's attack on an unbridled kind of liberal humanism, Halliwell and Mousley side with Soper's claim that anti-humanism tends to "secrete humanist rhetoric" (1986: 128). The almost proverbial discrepancy between the "earlier and middle" and "later" Foucault's position towards humanism is one of the examples they examine. If in *The Order of Things* and *The Archaeology of Knowledge* Foucault critiques humanism as an illusion of sovereignty and freedom, by the time he wrote "What is Enlightenment" (1984) this critique seems to be tempered by the recognition that the Enlightenment and humanism must be differentiated, and that "the humanistic thematic is in itself too supple, too diverse, too inconsistent to serve as an axis for reflection" (in Halliwell and Mousley 2003: 165). Furthermore, as this type of analysis usually argues, in the last two volumes of the uncompleted *The History of Sexuality* (1985, 1986), Foucault shifts his attention from the genealogy of modern social institutions and knowledges to an analysis of ancient Greek techniques of self-regulation and ethical self-production, that suggests traces of an aestheticized, Romantic notion of the self.[21]

What's more, Halliwell and Mosley argue that not only can humanist rhetoric be identified throughout anti-humanism, but that, as the flipside to this argument, humanism can also be perceived as secreting anti-humanist ideas; "humanism", they argue, "is always shadowed by its negation" (2003: 190). Similarly, Neil Badmington (2000, 2003) also argues that posthumanism already occurs as a critical practice within humanism. Drawing on a latent critique of Cartesian humanism within Descartes' work – namely, what he argues is an untenable distinction between human and machine in *Discourse on the Method* – Badmington seeks to theorize the posthuman not as a radical break with humanism, but as a "working through" of humanist assumptions, emphasizing what of humanism necessarily persists in posthumanism. Badmington finds his warrant in Derrida's "The Ends of Man" (1982),

---

[21] Many theorists view Foucault's later work as contradicting his earlier genealogies by expressing a kind of "nostalgia" for Enlightenment humanism. Of course, such an interpretation itself also depends on what kind of humanism this later "return" to humanism implies. Hardt and Negri, in line with their account of two traditions of humanism, offer a different reading of Foucault's work on technologies of the self, as an attempt to think humanism after the "death of Man", an attempt to articulate an "anti-humanist" humanism, or an attempt to bring to life the "creative life force" that animates the first revolutionary mode of immanent humanism:

> The humanism of Foucault's final works, then, should not be seen as contradictory to or even as a departure from the death of Man he proclaimed 20 years earlier. … This is humanism after the death of Man: … the continuous constituent project to create and re-create ourselves and our world (2000: 92).

It is in light of this understanding that Foucault's later works will be significant for the mediated posthumanist perspective that will be developed later on.

where, instead of proclaiming the "death of Man" like many of his contemporaries, Derrida countered that thought is never a "pure outside", but always takes place within a certain tradition and thus must bear some trace of it. Continuing Derrida, Badmington writes:

> Western philosophy is steeped in humanist assumptions ... the end of Man is bound to be written in the language of Man. ... To oppose humanism by claiming to have left it behind is to overlook the very way that opposition is articulated (2000: 9).

Humanism, then, is not an unambiguous concept. As these theorists observe, it is often reduced to only one of its many versions, and in its rich form, is already engaged with elements of anti- or posthumanism, thus challenging the very possibility of the latter. This could indeed render any definition of posthumanism more difficult to delimit than it has been until now. Yet, for all of humanism's "bagginess", the greater part of posthumanist discourse converses with only *one* account of humanism (albeit from different positions): that form of classical or liberal humanism that Soper has defined as appealing to the notion "of a core humanity or common essential features in terms of which human beings can be defined and understood" (1986: 11–12), based on a model of the human inherited from the Cartesian subject of the Cogito, the Kantian "community of rational beings" and the Lockean subject as property-owner and rights-holder. The humanism that the "post-" of posthumanism refers to then is the tradition which has upheld the subject as a free, autonomous, rational, self-contained and integral being that is detached in some fundamental way from the empirical world, unique and distinct by virtue of its being human. And this applies to both the biological understanding of the human as a bounded, "molar" organism, as shall be discussed in Chap. 5, and to the more theoretical understanding of the human as an autonomous, rational subject, a fully conscious self who possesses reason and a fixed internal identity that is independent from an outside world. As shall be argued, this humanist ontology always assumes a rigid separation of humans and "the rest" – without which there could be no claim for the uniqueness of the human – whether this be the "rest" of the natural world, populated by non-human living organisms, or the "rest" of the artificial world, populated by non-human things and technologies. It is vis-à-vis *this* type of humanism that our final axis of differentiation can be drawn, since it is this type of humanism that dystopic posthumanism is fearful of losing, that liberal posthumanism attempts to extend and that both radical and methodological posthumanism see as a discourse that must be overcome.

2.3.2 In Defence of Humanism

Dystopic posthumanism can be seen as an impassioned defense of humanism in these terms. It is predicated on the claim that there is something special about humans, something that entitles "every member of the human species to a higher moral status than the rest of the natural world" (Fukuyama 2002a: 166), and that

that something is under threat by advances in biotechnology. For dystopic posthumanists, this special something, whether it is called "human nature", "human dignity" or a "human essence", resides first and foremost in our existence as a biological species. For Fukuyama, it is "the species-typical characteristics shared by all human beings qua human beings" (2002a: 101). For Habermas (2003), it is located somewhere in the contingency of human birth, a "beginning which eludes human disposal" (58), that is, a state where only nature has a real determining influence and where one finds the existential human freedom to act. For Sandel (2007), similarly, an essential aspect of our humanness lies in the givenness or "giftedness" of life, i.e. in the random bestowing of talents and powers by nature that takes place at a time before the individual can willfully develop and exercise them. While this human essence or nature is never explicitly defined in the writings of dystopic posthumanism (an ambiguity that often becomes the focus of liberal posthumanist critique), it is nonetheless clear that it is something that all humans share by virtue of being human, that it is relatively unchanging and that it is unique to them. It is because biotechnologies interfere at this very basic biological level where human nature is taken to originally reside, because they can disrupt the boundary that encloses all humans in a single group, that they are such a threat, and a threat that grows proportionally to the degree of intervention (it is for this reason that genetic intervention, at the reproductive level in the form of PGD, or the idea of germline engineering, are particularly disputed). This conception of the human and human nature is predicated on the humanist dualist ontology. A clear line can be drawn between what is natural and what is unnatural and between subjects and objects. The latter are open to forms of manipulation that are unacceptable to the former.

What's more, these theorists often ascribe to the claim that the existence of a constant and identifiable human nature is what defines the scope of our political and social order; that the values, rights and moral sense that are written into our democratic and constitutional arrangements are shaped and constrained by the acknowledged membership in a biological species that serves as the basis for mutual respect. The unique, unchanging essence that distinguishes humans from non-humans at the biological level is translated here onto the politico-legal level, where human subjects are respected as autonomous, free and self-defining beings: "much of our political world," Fukuyama writes, "rests on the existence of a stable human 'essence' with which we are endowed by nature" (217), and for Habermas it is the "species ethic" that underwrites our self-understanding as the undivided, autonomous authors of our own lives. Thus the "integrity" and the "inviolability" of the human that humanism presupposes is upheld here on both the biological and the politico-legal levels, and needs to be defended from biotechnologies that insidiously cross profound ontological divides.

It is interesting, then, that transhumanism also sees itself as a champion of humanist values. As we shall see in the next chapter, this discrepancy arises in part from differing interpretations by dystopic and liberal posthumanism of what the translation of liberal humanist values should be in political or legal terms. But more importantly it is because both approaches ascribe to a very similar model of human nature based in the humanist distinction between humans and non-humans. In liberal posthumanist discourse, posthumanism does not indicate a break from humanism, but an extension of humanist ideals into a posthuman era. In these accounts the essential, rational self endures unimpeded – augmented and improved, certainly, but not essentially transformed by technological enhancements. Liberal posthumanism can be seen as the self-declared heir of classical secular humanism, its belief in the prospects of science and its vision of humanity freed from the constraints of superstition, ignorance and fear. Indeed, transhumanism models itself as a type of humanism, which can be viewed, declares Bostrom, as "an outgrowth of secular humanism and the Enlightenment" (2003); while in Max More's "Extropian Principles" we find that,

> like humanists, transhumanists favor reason, progress, and values centered on our well being rather than on an external religious authority. Transhumanists *take humanism further* by challenging human limits by means of science and technology combined with critical and creative thinking. (1998, Version 3.0, emphasis added)

This extension or continuation of humanism is manifest in three recurrent themes of liberal posthumanism: the project of controlling and transcending nature, a hyper-individualism and a Cartesian mind/body dualism.

The promise that emerging bio- and enhancement technologies hold for liberal posthumanism is of a liberation from the human species' biological bondage to finitude, disease and decay. Rising above these limitations presupposes that humans are already in some fundamental way distinct from their natural environment, and can be viewed in this sense as a deepening of the humanist ontological divide. In the first instance, this posthumanist liberation involves a mastery of the natural environment, depicted as a potentially hostile environment that, with its faulty and fickle mechanisms and its overwhelming arbitrariness, too often constitutes a main obstacle in the human's will to live. Liberal posthumanism can be seen in this sense as participating in the anthropocentric project of modernity as a harnessing and colonizing of nature. Furthermore, the transcendence of nature is often framed as part of a project of individual freedom, self-improvement and "personal growth" (as stated in an earlier version of the Transhumanist Declaration) that extends the narrative of the self-determining subject into "The dawn of a new humanism", thus baptized in the editorial of the first issue of *Mondo 2000* magazine.

But mastery over nature is not only a question of controlling the external natural environment. It also involves a mastery of nature *within*, as the natural human body, with *its* precarious, defective and unruly ways. This can involve replacing parts of the natural body that are failing or can be improved with artificial, human-made parts (cochlear implants, pacemakers, prosthetic limbs), or ultimately, discarding the organic body altogether by uploading the mind to a computer. If mind uploading is a somewhat radical take on enhancement, though a common one in transhumanist writings, it is nonetheless a logical step in the aim of complete independence from nature that is sought for via human enhancement. It is an extreme version of the traditional mind/body or form/matter dualism, where mind, conceived of as an informational pattern or code, is easily detached from the biological substrate of the organic body and relocated to a more robust and controllable one, but where the sense of a clearly defined, autonomous self in the humanist sense does not get lost in the move. As Moravec writes in his evocative description:

> Bit by bit our failing brain may be replaced by superior electronic equivalents, leaving our personality and thoughts clearer than ever, though, in time, no vestige of our original body or brain remains. (1999: 169–170)

2.3.3   The Cybernetic Human and Cybernetic Humanism

It is clear then, that the "post-" in liberal posthumanism does not refer to a rupture with the tradition and ideals of humanism, quite the opposite, it attempts to engrave these in a new technological/evolutionary setting. Liberal and dystopic posthumanism differ here. Both approaches presuppose the humanist categorical separation between humans and their environment, but while this model is a basis for both a biological and a more discursive or philosophical understanding of the human for dystopic posthumanism, this is not the case for liberal posthumanism, which subscribes to a different understanding of the human in the biological sense. As we have seen, dystopic posthumanists, even if they do acknowledge a certain malleability of human nature, argue for important, underlying biological elements that are unchanging, that constitute "a safe harbor" in the words of Fukuyama (2002a: 218) from which humans can identify, connect and respect one another. In liberal posthumanist discourse, on the other hand, the most significant property of the human as a biological organism is its transformative and dynamic character. Liberal posthumanism ascribes to a neo-Darwinian framework by which humans have and are constantly evolving via the mechanism of natural selection; they affect and are affected by their environments. The human in its current state is thus neither a cosmic given nor a stable and perfected final phase in evolution, but a particular phase of ongoing evolutionary transformation. And the current, particular phase of evolution is an unprecedented one insofar as evolution can now be driven or guided by humans themselves.

In the transhumanist narrative of evolution, this phase of "participant evolution", or the ability to steer evolution through scientific means, is a novelty, but it is also part and parcel of the evolutionary logic of survival and adaptation itself.[22] That is, it is the next "natural", evolutionary step for humans, who have always attempted to improve their chances of survival by gaining some degree of control or influence over their environment, namely through technology. This is to say that the "post-" in liberal posthumanism refers to a break that takes place within the evolutionary, biological conception of the human – but only in a rather narrow sense of "the human" as the product of one specific phase of human evolution, in which humans did not consciously steer evolution via technological means. It does not, however, refer to a break with humanism, as illustrated above. Liberal posthumanism is thus

---

[22] The idea of participant evolution, of deliberately redesigning the human, was first put forward by Clynes and Kline (1995).

specifically post*human*, and not post*humanist*. Not only because it argues straightforwardly for a continuation of humanist values, but because it maintains and deepens the humanist ontological divide between humans and their environment by arguing that humans can and do transcend and master nature. Stated differently, the dynamic or adaptive nature of human organisms to affect and be affected by their environments does not seem to pose a problem for the idea of the human's transcendent position vis-à-vis that environment. The human subject, though not the human body, remains here a humanist one, autonomous, bounded, fixed. This capacity to argue for a post-*human*-ism without what should be its correlate post-*human-ism* gives rise to a paradox.

It is helpful here to take a closer look at liberal posthumanism's model of the human qua biological organism not only in terms of Darwinian evolution, but also in terms of cybernetics. The cybernetic model of systems that gained influence in the second part of the twentieth century was based in the idea that living and mechanical systems alike depend on the processing of negative and postitive feedback, or circularity and recursivity (Hayles 1999; Heims 1991). The underlying premise of cybernetics was a radically new theory of information, that construed information as a purely quantitative or probabilistic entity that could be distinguished from the material or physical channel or substrate that carried it.[23] Essentially, this meant that both machines and living organisms could be recast in terms of information, i.e. that one single explanatory model could be used for biological and non-biological entities.

The notion of the feedback loop is crucial here. For cybernetics, information flows *through* networks, running from the environment through a system (either organic body or machine), so that it no longer makes sense to distinguish where elements of a single system begin or end. Gregory Bateson (1972) famously illustrated this with the classical example of a blind man using a stick, asking what the boundaries of the blind man's "system" are, and if it does or does not include the stick. The blind man's system, argued Bateson, does not end at his brain stem, nor at his fingertips, nor even at the tip of his cane. His arm, hand, cane, curb and ear all form a self-corrective cybernetic circuit that extends beyond the body. It is this cybernetic model of the human that informs liberal posthumanism, in which technology is not simply viewed as an extension or an appendage to the human body, but is often incorporated and assimilated into its very structures. In many transhumanist accounts at least, bodies are often immersed in technologies or technologies are wholly incorporated into the body, as in the examples of cognitive enhancement,

---

[23] Information became a mathematically defined concept thanks to theorists like Norbert Wiener and Claude Shannon. Shannon's *The Mathematical Theory of Communication* (1949) was a breakthrough in information theory in which he accomplished two radical and related moves. First, in Shannon's interpretation of communication as the transmission of a message from an information source (sender) along a channel to a destination (receiver), communication was defined in terms of the *selection* from a set of *possible* messages. Secondly, as a function of possibilities, information could then be disconnected from semantic issues – if information was tied to meaning, its values would have to change with every new context. The result was a definition of information as a probability function that has no materiality and no necessary connection to meaning.

wearable computers or mind uploading. Hence, even in their most extreme form, accounts of "post-biological" futures like Moravec's, where the human species may be declared obsolete, survival lies in the successful assimilation of humans and machines via uploading.

The cybernetic model thus contributes to a displacement of the unique position of the human by arguing for a conceptual absence of differentiation between humans and non-humans. Firstly because all systems, be they organic or non-organic, are construed in terms of information systems, and secondly because the interesting units of reference become networks – or posthumans – that are made up of organic and non-organic elements, systems and environments, through which information flows. But while this cybernetic turn does problematize the humanist ontological distinction between humans and their environments (natural or technological) on a biological level of understanding of the human, it does not problematize the humanist ontological distinction in terms of human selfhood and subjectivity. Some initial, unified self remains intact and essentially unpenetrated in liberal posthumanist accounts of emerging biotechnologies. The attitude of most transhumanists to the process of mind uploading mentioned earlier illustrates this.

Mind uploading assumes that there is something essential to selfhood which can be reduced to a computational configuration and can be relocated and preserved in a different medium – the computer – without the essential properties of the original configuration being significantly changed. Ray Kurzweil (2005), who has written quite extensively on mind uploading, calls this view "patternism", where the "patterns" of an individual mind, made up of things like the brain's sensory systems, the neural circuitry that makes up one's domain of general reasoning, one's attentional system, one's memories, etc., are what make up the self. The philosophical problems that patternism and mind uploading pose for notions of personal identity and selfhood are a heated topic of discussion among transhumanists: can a copy of a pattern be "me" if "I" am still "here"; if more than one copy of a pattern is made, am "I" also in the multiple copies? In order to preserve the unity of identity, is it better to upload all at once, or gradually, say neuron by neuron?[24] But while a number of transhumanists acknowledge that the implications for mind uploading on the notion of personal identity, often in very practical terms, are problematic, there is quite some agreement that an essence of selfhood can be maintained subsequent to uploading. Bostrom, in the Transhumanist FAQ, writes:

> Many philosophers who have studied the problem think that at least under some conditions, an upload of your brain would be you. A widely accepted position is that you survive so long as certain information patterns are conserved, such as your memories, values, attitudes, and emotional dispositions, and so long as there is causal continuity so that earlier stages of yourself help determine later stages of yourself. Views differ on the relative importance of these two criteria, but they can both be satisfied in the case of uploading. For the continuation of personhood, on this view, it matters little whether you are implemented on a silicon chip inside a computer or in that gray, cheesy lump inside your skull. (Bostrom 2003)

---

[24] For some important discussions around these questions and how different transhumanist thinkers deal with them, see Schneider (2009) and Hughes (2013).

Thus, even while the *body* is recognized as having a fundamentally dynamic and perhaps cyborg nature, the *self* or the *subject* continues to be understood in humanist terms.

This liberal posthumanist paradox is the same one N. Katherine Hayles (1999) identifies in the work of the early cyberneticians, in her tracing of the history, philosophy, and literary permutations of the movement. According to Hayles, the breakdown of boundaries implied in the notion of feedback, and the elevation of machines to the rank of complex intelligent entities alongside animals and humans in the framework of information, offered a radically new way of understanding human beings. So much so that she locates the beginning of the end of liberal humanism's "masterful subject of technology" with the emergence of the cybernetic paradigm. Nevertheless, Hayles argues, the early cyberneticians were committed to *preserving* rather than subverting humanist values such as autonomy and individuality, and they strived, consciously or not, to contain cybernetics within the circle of liberal humanist assumptions.

This was the source of a constant tension, Hayles explains, that is felt throughout the writings of cybernetics' founding father Norbert Wiener, as the attempt to reconcile his work in envisioning new ways of equating humans and machines and his belief in liberal humanist values. Indeed, while the concept of feedback provides the basis for the theoretical elimination of the frontier between the organic and the non-organic and for a cybernetic classification of beings along a spectrum of complexity, Wiener was not so much interested in showing that man was a machine as much as demonstrating that machines could function like men – in the image of an autonomous, self-directed individual. Hayles writes,

> placed alongside his human brother … the cybernetic machine was to be designed so that it did not threaten the autonomous, self-regulating subject of liberal humanism. On the contrary, it was to extend that self into the realm of the machine. (1999: 86)

Similarly, it seems that liberal posthumanists do not grasp the full extent of the ontological implications that a cybernetic epistemology entails. In their worldview, technologies enhance, but never compromise the unique position of humans in their natural and technological environments, and the ontological divide between humans and the rest of the world is maintained.

2.3.4   Non-Humanist Posthumanism

For radical and methodological posthumanism, posthumanism signifies a fundamental break with humanism, with the notion of the human that it champions, on both a biological as well as a philosophical or discursive level. Here the body *and* the idea of the human is opened up to its environment. In terms of cybernetics, one could say that the implications for the relational nature of being human that emerge from the shift in cybernetics research from the study of the internal specificity of objects to their interactions, are taken seriously here. Or, that radical and methodological posthumanists carry out the epistemological implications of the cybernetic paradigm to their logical (and ontological) conclusion.[25]

As we have already seen, for radical posthumanism new bio- and enhancement technologies are viewed as actively contributing to a deconstruction of foundational terms like "nature" and "the human", by physically collapsing the boundaries between the natural and the technological. The hybrid entities that emerge from these fusions and who are already part of our world, from slightly medically-enhanced humans through to full-blown cyborgs, are in this sense merely the actualization of the theoretical assertion, pronounced already by many precursors of radical posthumanism, that the human never was an integral, autonomous being exercising control over itself and its surroundings through individual agency and choice. This also means that even biologically unaltered humans can be considered posthuman; that, as Haraway claims, the cyborg *is* our ontology (1991a: 150), or that as Halberstam and Livingston write, "You're not human until you're posthuman" (1995: 8). The posthuman for radical posthumanism is thus not some literal "end" of the human, only of a certain image of it, and one that is more truthful to the experience of being human. In this understanding of posthumanism, any attempt to re-instate or resurrect a humanistic notion of human nature, as in the work of Fukuyama, is viewed as providing "fresh reinforcement for the crumbling humanist barricades in the rising tides of posthumanity" (Simon 2003: 3) – a desperate attempt to hold on to the qualities of autonomy and mastery put in place by liberal humanism, and a denial of what technology today really tells us about human nature.

While methodological posthumanism is often not as explicit or engaged about its position vis-à-vis humanism, its analyses of human-technology interactions also attempt to move beyond the human-centered framework of humanism and its human/non-human dualism. Actor-network theory (ANT), for example, argues that all elements in a network, whether these are humans, technologies, institutions or instruments, are "actants", that all play equally important roles in the construction of networks and their effects. This means that the "essence" of actants within a network (as subject, object, human or non-human) is bracketed off, in order to focus on how they engage, and that in any given network the identity of humans and non-humans is defined only through their interactions with other actors. ANT can be seen as a means of overcoming what is taken to be an unnecessary and unuseful duality between humans and non-humans, the result of what Latour (1993) has called modernity's "iconoclastic" shattering of the world into two ontologically separate categories. This ontological leveling of humans and non-humans, what is also known as "generalized symmetry" (Callon 1986), is found to various degrees in the works of methodological posthumanists. For Verbeek (2005, 2011), for example, the need for an ethics of technology that can take into consideration how technologies bring about situations that impose new moral requirements on human beings calls us to move beyond the humanist distinction between subjects and objects. Only such a "posthumanist ethics", one that does not take this distinction as its starting

---

[25] See Jean-Pierre Dupuy (1994) and Céline Lafontaine (2007) for good discussions about the similarities between French postmodern theory and cybernetics, which also make this point.

point, can account for the co-constitutive nature of morality as that which emerges at the meeting point between humans and non-humans, and act as a guide to living better with our contemporary technologies.

As I have suggested earlier, and will further develop, the important though somewhat subtle difference between methodological and radical posthumanism is mainly in the breadth of their aspirations. Methodological posthumanism can be seen as more descriptive then prescriptive, as aiming to develop better toolkits for analyzing relationality and materiality, then to develop foundational, explanatory accounts (Law 2010). Radical posthumanism is brazenly prescriptive. It has a clear posthumanistic political agenda, to "destabilize the ontological purity of Western modernity" (Graham 2002: 16). In this approach, if humanism has been detrimental to human and non-human forms of alterity, than an engagement with posthuman technologies offers an opportunity to contend for a different, more ethical vision of the human and those values that always accompany such visions. It is this future that is at stake, the foreclosure or cooptation of which by liberal posthumanist visions must be prevented. Hayles writes,

> Whereas the "human" has since the Enlightenment been associated with rationality, free will, autonomy and a celebration of consciousness as the seat of identity, the posthuman in its more nefarious forms is construed as an informational pattern that happens to be instantiated in a biological substrate. There are, however, more benign forms of the posthuman that can serve as effective counterbalances to the liberal humanist subject, transforming untrammeled free will into a recognition that agency is always relational and distributed, and correcting an over-emphasis on consciousness to a more accurate view of cognition as embodied throughout human flesh and extended into the social and technological environment. (2006: 160–161)

Nonetheless, both approaches take as their starting point the need to move beyond humanism, insofar as the categorical distinction upon which humanism is predicated, between an autonomous human subject and an objective world that it seeks to master, is a hindrance to understanding the many intricate ways in which humans and non-humans are actually interwoven, and how the experience of being human is always shaped by this interaction.

2.4   Conclusion

From this preliminary attempt to map the notion of the posthuman through various axes, the pessimist/optimist, historicist-materialist/philosophical-ontological and humanist/post-humanist, four types of posthumanist discourse come into view. Dystopic posthumanism argues against the use of technology to modify or enhance humans and views the posthuman as an imminent crisis that can have disturbing or dire repercussions on the fundamental meaning of human nature and the social values that are grounded in it. Dystopic posthumanism gravitates towards the pessimist, historicist-materialist and humanist poles of our axes. Liberal posthumanism advocates the desirability of improving the human condition by using new technologies to transcend the biological limits of the human species and views the right to use enhancement technologies as an extension of individual freedom and choice. Liberal posthumanism gravitates towards the optimist, historicist-materialist and humanist poles.

It is these two approaches to posthumanism that most commonly frame the public debate on posthuman and enhancement technologies. But while they do differ greatly insofar as dystopic posthumanism fervently objects to the use of enhancement technologies and liberal posthumanism welcomes them, both these approaches are grounded in the humanist view that humans are independent, autonomous entities that are clearly demarcated from their environments and from their technologies. This is true of liberal posthumanism even if it adopts a cybernetic model of the human on the biological level. It is because of this common humanist framing that these approaches, as I will argue throughout this research, are inadequate for grasping the real novelty of emerging biotechnologies and for capturing the complexity of the posthuman age – because it is precisely the humanist division of the world into subjects and objects, humans and non-humans, that these technologies so overtly undermine. It thus becomes clear that the optimist/pessimist axis which underlies the discussion about posthuman technologies in its most popular forms is of much less significance than the humanist/non-humanist axis.

The next approach that emerged from this mapping is radical posthumanism, which views the idea of the co-evolution of humans and technology as liberating from the humanist notion that the human is a fixed category that has been historically defined in opposition to its constitutive others. Informed by the anti-humanism of poststructuralist thought, radical posthumanism views the posthuman as providing a means of political resistance against the metanarratives of modernity and of offering a better account of the human as an ongoing process of technological and anthropological evolution. Radical posthumanism gravitates towards the optimistic, both the historicist-materialist and philosophical-ontological, and non-humanist poles of our axes. I designate this approach radical because it calls for a radical rethinking of human ontology in light of emerging biotechnologies. Finally, methodological posthumanism, like radical posthumanism, argues that the experience of being human is always shaped by our interactions with technology, that humans are "technologically mediated". I designate this approach methodological because of its emphasis on the need to develop new conceptual tools to analyze these interactions. Methodological posthumanism tends be neutral in terms of optimism or pessimism regarding emerging biotechnologies, and gravitates towards the historicist-materialist pole and the non-humanist pole of our axes.

For radical and methodological posthumanism the rigid separation of the world into humans and non-humans undertaken by modern metaphysics is only one possible configuration of the relations between humans and technologies, an understanding that allows for the creation of other configurations that can take into account the multiple ways in which humans and non-humans are interwoven. This holds for both the biological and the philosophical understanding of human being. It is because of this assumption that radical and methodological posthumanism offer much-needed alternatives to dystopic and liberal posthumanism, and better means of understanding and assessing the implications emerging biotechnologies have for what it means to be human. Chapters 5 and 6 explore these approaches and shed light on the limitations that they too present, before developing another perspective, "mediated posthumanism", that I suggest can integrate the important insights of both these approaches while overcoming their shortcomings. It is this final approach, I will argue, that can best account for the implications of these technologies for what it means to be human and perhaps for an emerging ethics of emerging biotechnology.

References

Achterhuis, H. (2001). *American philosophy of technology: the empirical turn.* Bloomington: Indiana University.

Agar, N. (2004). *Liberal eugenics: In defence of human enhancement.* Oxford: Blackwell.

Agar, N. (2010). *Humanity's end: Why we should reject radical enhancement.* Cambridge, MA: MIT Press.

Akrich, M. (1992). The de-scription of technical objects. In W. Bijker & J. Law (Eds.), *Shaping technology/building society* (pp. 205–224). Cambridge, MA: MIT Press.

Althusser, L. (1992). Ideology and ideological state apparatuses. In A. Easthope & K. McGowan (Eds.), *A critical and cultural theory reader.* Toronto: Toronto University Press.

Annas, G. (2005). *American bioethics: Crossing human rights and health boundaries.* Oxford: Oxford University Press.

Annas, G., Andrews, L., & Isasi, R. (2002). Protecting the endangered human: Toward an international treaty prohibiting cloning and inheritable alterations. *American Journal of Law and Medicine, 28*(2&3), 151–178.

Badmington, N. (Ed.). (2000). *Posthumanism.* New York: Palgrave.

Badmington, N. (2003). Theorizing posthumanism. *Cultural Critique, 53,* 10–27.

Badmington, N. (2004). Post, oblique, human. *Theology and Sexuality, 10*(2), 56–64.

Balsamo, A. (1996). *Technologies of the gendered body: Reading cyborg women.* Durham/London: Duke University Press.

Bateson, G. (1972). *Steps to an ecology of mind.* New York: Ballantine Books.

Bernal, J. D. (1929). *The world, the flesh and the devil: an inquiry into the future of the three enemies of the rational soul.* London: Jonathan Cape.

Bhaba, H. (1994). *The location of culture.* London: Routledge.

Bijker, W. E., Hughes, T. P., & Pinch, T. (Eds.). (1987). *The social construction of technological systems: new directions in the sociology and history of technology.* Cambridge: MIT.

Bordo, S. (1993). *Unbearable weight: Feminism, Western culture, and the body.* Berkeley: University of California Press.

Bostrom, N. (2003). The Transhumanist FAQ, Version 2.1. http://www.transhumanism.org/resources/FAQv21.pdf. Accessed 13 June 2013.

Bostrom, N. (2005). In defense of posthuman dignity. *Bioethics, 19*(3), 202–214.

Bostrom, N. (2009). Why I want to be a posthuman when I grow up. In B. Gordijn & R. Chadwick (Eds.), *Medical enhancement and posthumanity* (pp. 107–137). Heidelberg: Springer.

Braidotti, R. (2002). *Metamorphoses: Towards a materialist theory of becoming.* Cambridge: Polity Press.

Braidotti, R. (2006). *Transpositions: On nomadic ethics.* Cambridge: Polity Press.

Buchanan, A. (2011a). *Better than human: The promise and perils of enhancing ourselves.* New York: Oxford University Press.

Buchanan, A. (2011b). *Beyond humanity? The ethics of biomedical enhancement*. Oxford: Oxford University Press.

Bukatman, S. (1993). *Terminal identity: The virtual subject in postmodern science fiction*. Durham: Duke University Press.

Butler, J. (1999). *Gender trouble: Feminism and the subversion of identity*. New York: Routledge. Original edition, 1990.

Callon, M. (1986). Some elements of a sociology of translation: Domestication of the scallops and the fishermen of St Brieuc Bay. In J. Law (Ed.), *Power, action and belief: A new sociology of knowledge* (pp. 196–233). London: Routledge & Paul Keagan.

Callon, M., & Law, J. (1997). After the individual in society: Lessons on collectivity from science, technology and society. *Canadian Journal of Sociology, 22*(2), 165–182.

Clynes, M. E., & Kline, N. S. (1995). Cyborgs and space. In C. Hables Gray (Ed.), *The cyborg handbook* (pp. 29–43). London: Routledge.

Davies, T. (1997). *Humanism*. London/New York: Routledge.

Deleuze, G., & Guattari F. (1977). *Anti-Oedipus: Capitalism and schizophrenia* (trans: Seem, M., Lane, H. R., & Hurley, R). New York: Viking Press. Original edition, 1972.

Deleuze, G., & Guattari, F. (1987). *A thousand plateaus: Capitalism and schizophrenia* (trans Massumi, B.). Minneapolis: University of Minnesota Press. Original edition, 1980.

Derrida, J. (1976). *Of Grammatology* (trans: Spivak, C. G.). Baltimore: Johns Hopkins University Press. Original edition, 1967.

Derrida, J. (1982). The ends of man. In J. Derrida (Ed.), *Margins of philosophy*. London: Harvester.

Derrida, J. (1989). *Of spirit: Heidegger and the question* (trans: Bennington, G., & Bowlby, R.). Chicago: Chicago University Press.

Dery, M. (Ed.). (1994). *Flame wars: The discourse of cyberculture*. Durham: Duke University Press.

Dery, M. (1996). *Escape velocity: Cyberculture at the end of the century*. New York: Grove.

Dupuy, J. P. (1994). *Aux origines des sciences cognitives*. Paris: La Découverte.

Ellul, J. (1965). *The technological society*. New York: Vintage.

Fausto-Sterling, A. (2000). *Sexing the body: gender politics and the construction of sexuality*. New York: Basic Books.

Featherstone, M., & Burrows, R. (Eds.). (1995). *Cyberspace/cyberbodies/cyberpunk: Cultures of technological embodiment*. London: Sage.

Foucault, M. (1984). What is enlightenment? In P. Rabinow (Ed.), *The Foucault reader* (pp. 32–50). New York: Penguin.

Foucault, M. (1985). *A History of sexuality, vol. 2: The use of pleasure* (trans: Hurley, R.). New York: Pantheon. Original edition, 1984.

Foucault, M. (1986). *A History of sexuality, vol. 3: The care of the self* (trans: Hurley, R.). New York: Pantheon. Original edition, 1984.

Foucault, M. (1989). *The order of things: An archaeology of the human sciences*. New York: Pantheon. Original edition, 1966.

Fukuyama, F. (2002a). *Our posthuman future: Consequences of the biotechnology revolution*. New York: Farrar, Straus and Giroux.

Fukuyama, F. (2002b). How to regulate science. *The Public Interest, 146*(Winter), 3–22.

Fukuyama, F. (2004). The world's most dangerous idea. *Foreign Policy, 144*, 32–33.

Glover, J. (2006). *Choosing children: The ethical dilemmas of genetic intervention*. Oxford: Clarendon.

Graham, E. L. (2002). *Representations of the post/human: Monsters, aliens and others in popular culture*. New Brunswick: Rutgers University Press.

Gray, C. H. (Ed.). (1995). *The cyborg handbook*. New York: Routledge.

Grosz, E. (1994). *Volatile bodies: Towards a corporeal feminism*. Bloomington: Indiana University.

Habermas, J. (2003). *The future of human nature*. Cambridge: Polity.

Halberstam, J., & Livingstone, I. (Eds.). (1995). *Posthuman bodies*. Bloomington: Indiana University Press.

Halliwell, M., & Mousley, A. (2003). *Critical humanisms: Humanist/anti-humanist dialogues.* Edinburgh: Edinburgh University Press.

Haraway, D. (1991a). A cyborg manifesto: Science, technology, and socialist-feminism in the late twentieth century. In D. Haraway (Ed.), *Simians, cyborgs and women: The reinvention of nature* (pp. 149–181). New York: Routledge.

Haraway, D. (1991b). *Simians, cyborgs and women: The reinvention of nature.* New York: Routledge.

Haraway, D. (1997). *Modest_Witness@Second_Millenium. FemaleMan©_Meets_Oncomouse™: Feminism and technoscience.* New York: Routledge.

Hardt, M., & Negri, A. (2000). *Empire.* Cambridge, MA: Harvard University Press.

Harris, J. (2007). *Enhancing evolution: The ethical case for making better people.* Princeton: Princeton University Press.

Hayles, N. K. (1999). *How we became posthuman: Virtual bodies in cybernetics, literature and informatics.* Chicago: University of Chicago Press.

Hayles, N. K. (2006). Unfinished work: From cyborg to cognisphere. *Theory Culture and Society, 23*(7–8), 159–166.

Heidegger, M. (1977). The question concerning technology. In D. Farell Krell (Ed.), *Martin Heidegger: Basic writings* (pp. 287–317). New York: Harper & Row.

Heims, S. J. (1991). *The cybernetics group, 1946–1953: Constructing a social science for postwar America.* Cambridge, MA: MIT Press.

Herbrechter, S., & Callus, I. (2003). What's wrong with posthumanism? *Rhizomes, Fall 7* http://www.rhizomes.net/issue7/callus.htm.

Hughes, J. J. (2004). *Citizen cyborg: Why democratic societies must respond to the redesigned human of the future.* Boulder: Westview Press.

Hughes, J. J. (2013). Transhumanism and personal identity. In M. Moore & N. Vita-More (Eds.), *The transhumanist reader* (pp. 227–233). London: Wiley-Blackwell.

Huxley, J. (1957). *New bottle for old wine.* London: Chatto & Windus.

Ihde, D. (1990). *Technology and the lifeworld: from garden to earth.* Bloomington: Indiana University.

Ihde, D. (1993). *Postphenomenology: Essays in the postmodern context.* Evanston: Northwestern University Press.

Ihde, D. (1998). *Expanding hermeneutics.* Evanston: Northwestern University.

Ihde, D. (2009). *Postphenomenology and technoscience.* Albany: State University of New York Press.

Irigaray, L. (1985). *This sex which is not one* (trans: Porter, C.). New York: Cornell. Original edition, 1977.

Jonas, H. (1979). *The imperative of responsibility: In search of ethics for the technological age.* Chicago: Chicago University Press.

Kaku, M. (1998). *Visions: How science will revolutionize the 21st century and beyond.* Oxford: Oxford University.

Kass, L. (1985). *Toward a more natural science: Biology and human affairs.* New York: The Free Press.

Kass, L. (1997). The wisdom of repugnance. *The New Republic, 216*(22), 17–26.

Kass, L. (2002). *Life, liberty, and defense of dignity: The challenge for bioethics.* San Francisco: Encounter Books.

Kass, L. (2003). Ageless bodies, happy souls: Biotechnology and the pursuit of perfection. *The New Atlantis, 1*(Spring), 9–28.

Kittler, F. (1999). *Gramophone, film, typewriter* (trans: Winthrop-Young, G., & Wutz, M.). Stanford: Stanford University Press.

Kurzweil, R. (1999). *The age of spiritual machines: When computers exceed human intelligence.* New York: Viking.

Kurzweil, R. (2005). *The singularity is near: When humans transcend biology.* New York: Viking.

Lacan, J. (1977). *Ecrits: A selection* (trans: Sheridan, A.). London: Tavistock. Original edition, 1966.

Lafontaine, C. (2007). The cybernetic matrix of 'French theory'. *Theory Culture and Society, 24*(5), 27–46.

Latour, B. (1992). Where are the missing masses? Sociology of a few mundane artifacts. In W. E. Bijker & J. Law (Eds.), *Shaping technology/building society: Studies in sociotechnological change* (pp. 225–259). Cambridge, MA: MIT Press.

Latour, B. (1993). *We have never been modern*. Cambridge, MA: Harvard University Press.

Latour, B. (1999). *Pandora's hope: Essays on the reality of science studies*. Cambridge, MA: Harvard University Press.

Law, J. (2010). The material of STS. In M. C. Beaudry & D. Hicks (Eds.), *The Oxford handbook of material culture studies* (pp. 171–186). Oxford: Oxford University Press.

Lévi-Strauss, C. (1963). *Structural anthropology* (trans: Jacobson, C., & Grundfest Schoepf, B.). New York: Basic Books. Original edition, 1958.

Locke, J. (1823). *An essay concerning human understanding*. In The works of John Locke. A New Edition, Corrected. In Ten Volumes, Vol. II, London, book 3, chapter 3.

Luhmann, N. (1995). *Social systems* (trans. Bednarz, J., & Baecker, D.). Stanford: Stanford University Press.

Lyotard, J. F. (1991). *The inhuman: Reflections on time* (trans: Bennington, G., & Bowlby, R.). New Haven: Yale University Press.

Lyotard, J. F., & Gruber, E. (1999). *The hyphen: between Judaism and Christianity* (trans: Brault, P. -A., & Naas, M.). New York: Humanity Books.

McKibben, B. (2003). *Enough: Staying human in an engineered age*. New York: Times Books.

Mitchell, R., & Thurtle, P. (Eds.). (2004). *Data made flesh: Embodying information*. New York: Routledge.

Moravec, H. (1990). *Mind children: The future of robot and human intelligence*. Cambridge, MA: Harvard University Press.

Moravec, H. (1999). *Robot: Mere machine to transcendent mind*. New York: Oxford University Press.

More, M. (1998). The extropian principles: A transhumanist declaration. http://www.maxmore. com/extprn3.htm. Accessed 12 June 2013.

Pepperell, R. (1995). *The post-human condition*. Oxford: Intellect.

Pickering, A. (1995). *The mangle of practice: Time, agency and science*. Chicago: University of Chicago Press.

Pickering, A. (2005). Asian eels and global warming: A posthuman perspective on society and the environment. *Ethics & The Environment, 10*(2), 29–43.

Rifkin, J. (1998). *The biotech century: Harnessing the gene and remaking the world*. New York: Tarcher/Putnam.

Roache, R., & Clarke, S. (2009). Bioconservatism, bioliberalism, and repugnance. *Monash Bioethics Review, 28*(1), 1–21.

Sandel, M. (2007). *The case against perfection: Ethics in the age of genetic engineering*. Cambridge: Harvard University Press.

Savulescu, J. (2001). Procreative beneficience: Why we should select the best children. *Bioethics, 15*(5), 413–426.

Savulescu, J. (2005). New breeds of humans: The moral obligation to enhance. *Ethics Law and Moral Philosophy of Reproductive Biomedicine, 1*(1), 36–39.

Savulescu, J. (2007). In defence of procreative beneficence. *Journal of Medical Ethics, 33*(5), 284–288.

Savulescu, J. (2010). Human liberation: Removing biological and psychological barriers to freedom. *Monash Bioethics Review, 29*(1), 4.1–4.18.

Schneider, S. (2009). Future minds: Transhumanism, cognitive enhancement and the nature of persons. In V. Ravitsky, A. Fiester, & A. L. Caplan (Eds.), *The Penn Center guide to bioethics* (pp. 844–856). New York: Springer.

Shannon, C., & Weaver, W. (1949). *The mathematical theory of communication*. Urbana: University of Illinois Press.

Simon, B. (2003). Toward a critique of posthuman futures. *Cultural Critique, 53*(Winter), 1–9.

Smith, W. J. (2004). *Consumer's guide to a brave new world.* San Francisco: Encounter Books.

Soper, K. (1986). *Humanism and anti-humanism.* London: Hutchinson.

Spivak, G. C. (1988). Can the subalterns speak? In C. Nelson & L. Grossberg (Eds.), *Marxism and the interpretation of culture.* Basingstoke: Macmillan.

Stapledon, O. (1931). *Last and first men.* London: Methuen & Co.

Stock, G. (2002). *Redesigning humans: Choosing our children's genes.* London: Profile.

Stone, A. R. (1991). Will the real body please stand up?: Boundary stories about virtual cultures. In J. Wolmark (Ed.), *Cybersexualities: A reader on feminist theory, cyborgs and cyberspace* (pp. 69–98). Edinburgh: Edinburgh University Press.

Stone, A. R. (1995). *The war of desire and technology at the close of the mechanical age.* Cambridge: MIT Press.

Swierstra, T., Boenink, M., & Stemerding, D. (2009). Exploring techno-moral change: The case of the obesity pill. In P. Sollie & M. Düwell (Eds.), *Evaluating new technologies: Methodological problems for the ethical assessment of technology developments* (pp. 119–138). Dordrecht: Springer.

Swierstra, T., van de Bovenkamp, H., & Trappenburg, M. (2010). Forging a fit between technology and morality: The Dutch debate on organ transplants. *Technology in Society, 32*(1), 55–64.

The President's Council on Bioethics. (2003). *Beyond therapy: Biotechnology and the pursuit of happiness.* New York: Regan Books.

Verbeek, P.-P. (2005). *What things do: Philosophical reflections on technology, agency and design.* University Park: Penn State University Press.

Verbeek, P.-P. (2011). *Moralizing technology: Understanding and designing the morality of things.* Chicago: Chicago University Press.

Vinge, V. (1993). The coming technological singularity: How to survive the post-human era. http://www-rohan.sdsu.edu/faculty/vinge/misc/singularity.html. Accessed 13 June 2013.

Waldby, C. (2000). *The visible human project: Informatic bodies and posthuman medicine.* London/New York: Routledge.

Warwick, K. (2002). *I, Cyborg.* London: Century.

Winner, L. (1980). Do artifacts have politics? *Daedalus, 109*, 121–136.

Wolfe, C. (2010). Posthumanities. http://www.carywolfe.com/post_about.html. Accessed 7 Mar 2011.

Wolmark, J. (Ed.). (1991). *Cybersexualities: A reader on feminist theory, cyborgs and cyberspace.* Edinburgh: Edinburgh University Press.

Zylinska, J. (Ed.). (2002). *The cyborg experiments: The extensions of the body in the media age.* London/New York: Continuum.

Chapter 3
The Human Enhancement Debate:
For, Against and from Human Nature

Abstract This chapter reviews the state of the ongoing debate between dystopic and liberal posthumanists on enhancement technologies, with a closer look at the explicit and implicit arguments advanced by each regarding some specific technologies like preimplantation genetic diagnosis, the use of psychopharmaceuticals for mood and cognitive enhancement, and genetic engineering. In broad terms, dystopic posthumanism subscribes to the moral claim that human enhancement is intrinsically wrong, and the political claim that it should be banned or restricted. Liberal posthumanism, conversely, holds that enhancement is neither intrinsically wrong nor unusually dangerous, and should generally be permitted. On both sides, the arguments that support these claims abound, and can be grouped into three categories: social, technical and methodological arguments.

Beyond these relatively commensurable terms, however, the debate between dystopic and liberal posthumanism is an ethical dispute at the core of which lie incommensurable views of human nature. While this is more obvious in the case of the dystopic posthumanist critique, which proceeds from the idea that technological intervention for enhancement purposes poses a threat to human nature, it is also the case that liberal posthumanism invokes human nature in its support of enhancement. Only, rather than extolling human nature as a fixed, stable and 'given' essence, it draws on a conception of the human as an evolving, dynamic and imperfect organism, who, by nature, aspires towards self-improvement.

Keywords Human enhancement • New eugenics • Human nature • Liberal posthumanism • Bioconservatism

Over the last decade, much of the reflection about notions of the posthuman and the impact of emerging biotechnologies on what it means to be human has centered around the debate on human enhancement, the use of medicine and technology to reshape, manipulate and enhance various aspects of human biology. In this context emerging biotechnologies are perceived as blurring the distinction between

T. Sharon, *Human Nature in an Age of Biotechnology: The Case for Mediated Posthumanism*, Philosophy of Engineering and Technology 14, DOI 10.1007/978-94-007-7554-1_3, © Springer Science+Business Media Dordrecht 2014

treatment and enhancement, and consequently raising fundamental questions concerning free will, human dignity and moral values. This chapter reviews the state of the ongoing debate between dystopic and liberal posthumanists on enhancement technologies, with a closer look at the explicit and implicit arguments advanced by each regarding some specific technologies like preimplantation genetic diagnosis, the use of psychopharmaceuticals for mood and cognitive enhancement and genetic engineering.

In broad terms, as we have seen, dystopic posthumanism subscribes to the moral claim that human enhancement is intrinsically wrong, and the political claim that it should be banned or restricted. Liberal posthumanism, conversely, holds that enhancement is neither intrinsically wrong nor unusually dangerous, and should generally be permitted. On both sides, the arguments that support these claims abound, and can be grouped into three categories: social, technical and methodological arguments. Beyond these relatively commensurable terms, however, the debate between dystopic and liberal posthumanism is an ethical dispute at the core of which lie incommensurable views of human nature. While this is more obvious in the case of the dystopic posthumanist critique, which proceeds from the idea that technological intervention for enhancement purposes poses a threat to human nature, it is, we shall see, also the case that liberal posthumanism invokes human nature in its support of enhancement. Only, rather than extolling human nature as a fixed, stable and 'given' essence, it draws on a conception of the human as an evolving, dynamic and imperfect organism, who, by nature, aspires towards self-improvement.

3.1 Treatment Versus Enhancement and New Designer Labels

The debate about emerging biotechnologies is regularly framed in terms of human enhancement, the notion that refers to the use of medicine and technology to improve one's physical and mental capacities beyond levels that are considered normal, and more specifically, the use of pharmacological agents, genetic engineering or biomedical implants, to enhance memory, intelligence, strength, endurance, agility or personality. Inherent to the notion of human enhancement is the idea – or the questioning of the idea – that it differs in some important way from treatment: that there is a fundamental distinction between *restorative* therapy and interventions that aim to bring about *improvements* extending beyond strictly therapeutic aims.

It is the moral quandary that arises when medical means are employed for non-medical ends, unrelated to curing or preventing disease, and when new biotechnologies prove too slippery for a rigid distinction between treatment and enhancement, that lies at the heart of the most heated discussions about emerging biotechnologies, and that has captured the attention of large-scale policy making, from the advisory committee on aspects of human enhancement recently established by the European Parliament, to George W. Bush's President's Council on Bioethics. The Council's widely read report, *Beyond Therapy: Biotechnology and the Pursuit of Happiness* (2003), for example, identifies four areas in which medical biotechnology is now moving "beyond therapy" to pursue goals of augmentation or transformation of life: (1) better children (prenatal diagnosis, embryo selection, genetic engineering of embryos, and behavior modification with drugs especially in relation to ADHD), (2) superior performance in sports, (3) ageless bodies (life extension technologies), and (4) happy souls (memory alteration and, in particular, mood improvement through the use of SSRIs).

Importantly, it is not so much the idea of self- and bodily improvement that opponents of human enhancement regard critically. Here they agree with pro-enhancement theorists that, at least to some extent, the attempt to increase one's health, fertility, lifespan, mental prowess, etc., has always been a part of what we humans do. But where current means of human enhancement differ greatly is first in their *technological* nature – and the perceived distinction between such technological forms of enhancement and "natural" ones – and second, in their *consumerist* nature – that is, the idea that such interventions are increasingly shaped by individual desires in a market culture.

These two concerns come together in the widespread use of the term "design", which has become one of the most powerful framing devices through which to interpret enhancement technologies, from designer babies to designer moods. As Sarah Franklin and Celia Roberts (2006) have noted in their work on preimplantation genetic diagnosis, the term "design" includes and unites a number of different meanings, with overlaps between design as something that is purposefully conceived or acts as a formulated plan ("a" design), design as desire (to have designs on something), and design as something that is tailor-made for a specific individual or elite (designer jeans). In the context of preimplantation genetic diagnosis, the term encompasses these different realms and phenomena, signifying what the authors call "a disturbing mixture of newfound biogenetic control, consumer demand and parental desire" (2006: 1). This vagueness of the term, they suggest, and its ubiquitousness, allows it to act as a "placeholder" for issues and opinions that are difficult to articulate. Interestingly, while design is employed as a very derogatory term by opponents of human enhancement, not all advocates of enhancement shun away from its use.[1] On the contrary, here design is desirable and even ethical. It is associated with the use of human reason and will to deal with the ageing, disease and other flaws that are part of nature, and with gaining some control over a fickle and indifferent evolutionary process that cares little for the fate of human beings.

Preimplantation genetic diagnosis (PGD) is a technique that was introduced in the 1990s that makes it possible to screen embryos for gene variants before transferring them to a woman's uterus. Embryos are produced using standard in vitro fertilization (IVF) procedures, biopsied and genetically screened in vitro. Only those embryos with the desired genetic profile are then selected and transferred to the uterus. Currently, there are mainly two groups of people who are using PGD:

---

[1] In a recent *Reader's Digest* article Julian Savulescu (2012), for example, provocatively declares that "it's our duty to have designer babies".

couples with a high risk of transmitting a single-gene disorder (such as cystic fibrosis or spinal muscular atrophy), and couples who have suffered a history of repeated miscarriage resulting from rare chromosomal translocations. More recent applications of PGD include diagnosing late-onset diseases and predisposition syndromes like cancer, and testing embryos for tissue matching so that they can later serve as cord blood or bone marrow donors to an existing affected sibling (what is known as "savior sibling"). Today genetic tests for more than 1,000 conditions are available, but technically speaking, virtually any genetic test that exists could be used in PGD.

While PGD was initially created as an alternative to prenatal genetic diagnosis that would allow parents to avoid having a child with a severe or deadly genetic disease, the technique is increasingly beginning to be used to target less-serious disorders or specific *traits*, such as an embryo's sex. This non-medical or so-called cosmetic application of PGD is an example of what dystopic posthumanists see as a boundary crossing from treatment to enhancement purposes, and it is this perceived potential for the deliberate selection of traits like eye color, athletic ability, height or intelligence that has branded it the "designer baby" technology. From a dystopic posthumanist perspective, this alleged use of PGD is viewed as the ultimate hubris of scientists and the symbol of too much genetic control released into the hands of choosy parents. As we shall see shortly, and as is characteristic of the human enhancement debate, this technology bears upon completely different values for dystopic or for liberal posthumanism. For dystopic posthumanism, such uses of PGD negates the Kantian-based value that children should always be appreciated as ends in themselves, as "gifts", rather than the products of our will or the instruments of our ambition. For liberal posthumanists the use of PGD, in both forms, is an extension of the right to procreative choice that should be left to the discretion of parents.

Similar to the designer baby idiom, the mood-enhancing effects of a number of psychopharmacological drugs, namely the SSRI family of anti-depressants like Prozac, are also interpreted through the frame of design. Prozac was originally approved for clinical depression and obsessive-compulsive disorder in the late 1980s. Owing to the drug's relatively few side effects, it quickly began to be prescribed for a wider range of mood-related ailments and became known for its ability to make people feel "better than well" (Elliot 2003). This use of psychopharmaceuticals to enhance mood and temperament in the absence of clear illness, has gained the use of drugs like Prozac the label of "cosmetic psychopharmacology" (Kramer 1993b). Like its counterpart the designer baby, this label assumes that the role of these drugs is not limited to the biological mechanisms of depression, but also might have an effect in shaping personality and the self, which, like genetic make-up at the embryonic stage, will become a matter of preference. More recently, the problematic use of antidepressants as mood enhancers has been extended to the realm of cognitive enhancement, the off-label use of neurological drugs such as Ritalin and Adderall (originally developed for ADHD) or Provogil (originally used for the treatment of narcolepsy and excessive sleepiness) to improve memory, concentration and planning.

3.2   Social and Technical Arguments

One general aspect of the human enhancement debate involves concerns about the safety and equality of access to enhancement technologies – issues that are familiar from discussions about most new technologies. From a dystopic posthumanist perspective the use of these technologies poses a number of risks that cannot be foreseen. In the debate on neuroenhancers, for example, it is argued that even if a drug has no immediate side-effects, harmful side-effects may appear 10 or 20 years later. A drug that improves short-term memory may have negative effects on long-term memory, or a drug that boosts mood may be habit-forming, with long-term effects on neurotransmitter function, the way amphetamines and opioids do. This precautionary principle holds that if we cannot predict the effects of enhancements, the default option should be to ban them, or to impose serious restrictions on their use beyond therapy. Furthermore, in determining the utility of enhancements, it is argued that not only benefits to the individual should be considered, but that these need to be weighed against the public good. Indeed, one of the main concerns dystopic posthumanists raise against the growing use of enhancement technologies is that they will change social relations by introducing new forms of inequality and discrimination.

Fukuyama (2002) for example, argues that the possibility of "buying" genes for one's children, which he believes will become one of the inevitable consequences of the new genetics and reproductive medicine, will have a disastrous social effect, aggravating existing inequalities by turning financial disadvantages into biological ones. In the future, he speculates, large gaps between the genetic haves and have-nots will make the cooperative relations characteristic of liberal societies unlikely. For Habermas (2003), one of the alarming effects of genetic enhancement is that a child who will have been the "product" of enhancement will never be able to confront his or her parents as a moral equal. This imbalance of power in the parent/child relationship will further be extended to an asymmetry in the moral community, where individuals will not recognize one another as moral equals.

From a liberal posthumanist perspective, the precautionary principle in itself cannot provide a working guideline because its consistent implementation would stifle all technological progress. Rather, questions of safety in the technological context require highly complex methods of risk assessment (Bostrom 2002). Another common argument advanced by liberal posthumanists is that much of the concerns expressed by opponents of enhancement are not based on an examination of the realities of contemporary science. While many of the phenomena of life currently seem to be understandable, we are still a long way from being able to re-engineer them at will, and despite the fact that biotechnology today, as many technologies, is clearly future-oriented and thrives on promises of epochal changes, liberal posthumanists often warn that bioethics is actually running far ahead of science, rather than the other way around. Thus, they often remark that the idea of being able to completely design one's child by selecting and inserting positive traits (rather than selecting for negative traits, which is what the prevalent use of PGD involves) results from both a misunderstanding of how genes function and how the technology works, and how costly and unpredictable it is. Steven Pinker has rather eloquently articulated this argument in response to Michael Sandel in the context of a talk given at Harvard in 2004:

> I think it's somewhat misleading to assume that parents will soon face the question, would you opt for a procedure that would give you a more talented or happier child? I think the real question is more likely to be something like, would you opt for a traumatic and expensive procedure that might give you a slightly more talented child, might give you a less talented child, might give you a deformed child, and might make no difference at all? I think that's more likely to be the choice, and it's not clear that hundreds of millions of people would say yes to it. (Pinker 2004)

Indeed, in light of the very random nature of genetic endowment, Pinker maintains that the highly critiqued notion of perfection in the framework of designer babies really is the "least of our worries".[2]

Regarding the aggravation of existing equalities, on the other hand, liberal posthumanists argue that in principle there are many ways in which this concern can be dealt with, by either making the technologies available to all, or even only the worst off, or via other compensating social policies (Hughes 2002). Furthermore, in an interesting twist, some liberal posthumanists contend that access to enhancement technologies can actually contribute to *alleviating* inequalities that arise from the natural, unequal distribution of biological or genetic capacities at birth, the effects of the so-called "genetic lottery", so that banning them could also be understood as perpetuating inequalities when technologies that could help reduce them exist.

This type of distributive justice argument is common in the discussion on mood-enhancing psychopharmaceuticals for example (Hughes 2009; Walker 2009), where liberal posthumanists take their cue from some findings in the burgeoning field of "happiness studies", aimed at the scientific understanding of subjective well-being. An important claim that has been made in this field is that subjective well-being, or what is known as the "happiness set-point", is in part determined by an initial brain setting that individuals are born with (Lykken and Tellegen 1996; Lykken 1999). Such studies, furthermore, claim that the happier people are in their original set-point, the more likely they are to succeed in areas that will make them happy, such as social achievements, better health, and more sought after jobs. This is to say that the capacity for happiness is unequally distributed at birth and then rewarded or penalized by society throughout life. Viewed in such terms, mood-enhancers can be a means of assisting people who have had the bad luck of being born with a low happiness set-point, and restricting access to them can be seen as unjust, and as a

---

[2] This argument is not only upheld by liberal posthumanists. Franklin and Roberts argue that what PGD involves from "up close" may be very different from what it seems from afar. Accounts of couples using PDG, they claim, are far removed from fashionable brave new world and designer baby anxieties. Rather, many of the couples they spoke with arrived at PGD as a last resort, following painfully traumatic experiences of watching young children die of terminal genetic disorders or long histories of repeated miscarriages: "Far from seeking offspring with genes for blond hair, blue eyes, [and] an imposing stature" they write, "…these parents, or would-be parents, simply want a child who will survive" (2006: 17–18).

hindrance to greater equality. In an article for *Free Inquiry* magazine, Mark Alan Walker explains:

> In terms of distributive justice, the creation of HPP [Happy-People Pills] is a no-brainer, for it would increase the pool of this very valuable resource. It would allow us to distribute it to those not lucky enough to win the genetic lottery for hyperthymia [above average happiness] without having to take anything away from anyone ... To deny the rest of us access to HPP is a grave form of injustice, for that would artificially limit the pool of this valuable resource ... [and] to prohibit most of us from the opportunity for what many (but not all) see as the best life: life with the happiness and achievement of the hyperthymic. (2009: 35–36)

But it is precisely the kind of social conformity that would accompany the widespread use of mood-enhancers like Prozac and cognitive-enhancers and stimulants like Ritalin and Adderall, producing an entire nation of hyperthymic individuals ready to tolerate and cope with the competitive realities of contemporary life, that is a main concern in the dystopic posthumanist approach. Enhancements are perceived as "a bid for compliance – a way of answering a competitive society's demand to improve our performance and perfect our nature" (Sandel 2004: 7), leading to a "drug-induced contentment" (Kass 2002: 48), which might thwart people's impulse to resist unjust, unrealistic or exploitative expectations and demands, and prevent wider social reform. In this context reference is often made to the "soma" drug of Aldous Huxley's *Brave New World*, that provides an easy escape from the ordeals of daily life and is employed by the government as a method of control through pleasure. For these theorists, psychopharmaceuticals enable a changing of our nature to fit the world rather than the other way around, diminishing our determination for social and political improvement.

One riposte to fears of social apathy by liberal posthumanists has been that "happy" people exhibit more pro-social behavior (Walker 2007). That on average, happier, more extrovert individuals are more assertive, prone to recognize and resist illegitimate authority, and have greater abilities to mobilize social networks (like conscious-raising groups or labor unions) than sad and introvert individuals. Bostrom has argued that cognitive and mood enhancement is actually a gateway to richer and more meaningful lives:

> Technologies such as brain-computer interfaces and neuropharmacology could amplify human intelligence, increase emotional well-being, improve our capacity for steady commitment to life projects or a loved one, and even multiply the range and richness of possible emotions. (2003: 5)

3.3 Eugenics Old and New

These issues of social justice and equality often crystallize in the framework of the discussion on eugenics. For dystopic posthumanist writers the debate on human enhancement often takes place in the shadow of the eugenic projects of the early twentieth century (Habermas 2003; Sandel 2004), where the use of genetic technologies, particularly human embryonic stem cell research and PGD to eliminate unwanted physical and possibly character traits, is associated with historical attempts in the quest for biological "improvement" through reproductive control.[3] Conversely pro-enhancement theorists argue that a crucial moral distinction can be drawn between the *authoritarian* and coercive eugenic programs epitomized by Nazi Germany – whose focus was the nation, race or class and where the state would impose eugenic choices – and a *liberal* eugenics, that focuses on the welfare of the individual, gives primacy to the individual's own values and conceptions of the good life, and where the role of state is limited to facilitating and enabling the choice of individuals for enhancement. For proponents of this "new" eugenics, the horrors associated with the coercive practices of the old eugenics should not blind us from the potential that new biotechnologies have for improving human welfare, and as long as the choice of whether to use enhancement technologies and how is left to individuals, human rights, pluralism and personal welfare will not be jeopardized.

The "distinguishing mark" of the new eugenics, in this sense, as Nicholas Agar writes in his book *Liberal Eugenics* (2004), is the neutrality of the state vis-à-vis the different conceptions people have of the good life and what characteristics are desirable. Just as a liberal perspective rules out any form of authoritarian eugenics where the state aims to mold humans according to its own particular views of fit and unfit, a liberal perspective encourages us to adopt a liberal eugenics in which parents are free to choose some characteristics of offspring based on their personal conceptions of human excellence. One of the arguments Agar makes to support this claim is a "nature principle": "If we are permitted to leave unchanged a given genetic arrangement in the genomes of future children, we are also permitted to introduce it" (2004: 99). That is, if some genetic arrangement is not considered the source of an impairment that would prompt us to intervene and change it (such as blue eyes, or height), then there is no moral reason precluding us from bringing that arrangement about. This means, Agar suggests, that the only restriction on genetic enhancement is the possibility of harm to the resulting child, in other words, the autonomy or freedom of that child. Buchanan, Brock, Daniels and Winkler also defend a similar type of liberal eugenics in *From Chance to Choice* (2000), where reproductive freedom is defended so long as the benefits and burdens of genetic improvement are "fairly distributed".

Some liberal posthumanist theorists push the endorsement of liberal eugenic practices further, arguing that human enhancement practices should not only be tolerated, but encouraged. John Harris (2007, 1993) argues that a concern for the welfare of the future of the human race implies that we have an obligation to pursue enhancements. He upholds that there is nothing morally wrong per se in practicing

---

[3] Habermas' *The Future of Human Nature*, with an original subtitle in German of "On the Way to a Liberal Eugenics?", is a specific contribution to this argument. For obvious historical reasons, the discussion on new biotechnologies in Germany, especially prenatal "selection" and euthanasia for severely disabled newborn infants, is particularly sensitive. And Germany has taken a very conservative attitude towards biotech practices that may be associated with eugenics, such as banning PGD.

eugenics, but that there is a wrong practice: which occurs the moment a majority restricts the reproduction of a "genetically weak" minority. For Harris, a liberal eugenics does not deny the "genetically weak" the right to reproduce, what it does is allow them to produce healthier children. He writes: "It is not that the genetically weak should be discouraged from reproducing but that everyone should be discouraged from reproducing children who will be significantly harmed by their genetic constitution" (1993: 183). In a number of works, Julian Savulescu has also argued that parents have a moral duty to use selective technologies to produce the "best children possible" (2008, 2007, 2005, 2001). His principle of "procreative beneficence" holds that:

> Couples (or single reproducers) should select the child, of the possible children they could have, who is expected to have the best life, or at least as good a life as the others, based on the relevant, available information (2001: 415).

More recently, Savulescu has argued that this principle also extends to moral enhancement (2008), arguing that if superior moral traits are genetically determined and can be screened for (as he believes that advances in genetics have shown), then people have a moral obligation to select "ethically better" embryos.

What was morally objectionable about the old eugenics, in this view, was its compulsory nature. The "new" eugenics, however, is voluntary; it is about enhancing people's freedom rather than reducing it. And as such, it opens up a space for diversity, pluralism and even experimentation, by enabling and facilitating the expression of the individual and particular values of human beings. Concerns of homogenization and conformity dissipate when the state withdraws. Agar writes:

> There will be no directive to evolve people towards a single optimal type. Rather, access to information about the full range of genetic therapies will allow the value of prospective parents to inform their eugenic plans. Differing ideas about the good life will surely disrupt any centrally directed pattern of enhancement. (2004: 146)

Similarly, Andy Miah (2009), with strong faith in humanity's aspiration and admiration of human variance ("as the history of fashion reveals") and pushing the market logic to its extreme, argues that "Once we have access to the full range of human modifications … instead of converging around a single notion of beauty, we will invent new forms of human beauty".

These arguments for pluralism, freedom and individual welfare do very little in the direction of assuaging the concerns of opponents of enhancement. For these theorists, the notion of coercion is a lot less clear-cut than it is for advocates of enhancement.[4] While various types of involuntary medical intervention are perhaps the most morally objective forms of eugenic practices, it is suggested that disciplinary pressures grounded in social norms and mediated by the market yield very similar eugenic effects. Enhancement, whether we call it the "new", "liberal", "privatized", "free-market" or "yuppie" eugenics (Hubbard and Newman 2002), differs little

---

[4] It is of some interest that the opponents of enhancement are the ones who call for a much more thorough analytical discussion of the complex underpinnings of the concept of coercion – not its advocates, for whom it plays such a pivotal role in the moral justification of enhancement.

from the "old" eugenics, insofar as both, writes Sandel, "make children into products of deliberate design". In this privatized version of eugenics, it will be a question of who can afford enhancement. This will lead to a deepening of the gap between rich and poor, a growing intolerance towards human capacities and morphologies that will differ from those defined by that financial elite, and the absence of choice on the part of parents concerning the use of enhancement, if they will want their children to be able to compete with "enhanced" children in the future. In this sense, talk of autonomy in the context of liberal eugenics is disturbingly misplaced.[5]

3.4   The Argument from Human Nature: Dystopic Posthumanism

From the dystopic posthumanist perspective then, eugenics is objectionable on other grounds, even when coercion is not involved. Namely, there is something wrong per se with the effort to select genetic qualities, to try to control or exercise dominion over the genetic traits of the next generation; and this is because it threatens something intrinsically valuable, more valuable than any of the benefits that science and technology promise to deliver: human nature. Human nature, here, and "the natural" in general, is deployed here as a moral category. That which is natural is morally valuable, and that which is unnatural is morally dubious. This distinction builds on and adds moral flesh to the very possibility of distinguishing between the natural and the unnatural, or the human and "the rest", that lies at the heart of the humanist narrative, as I argued in the previous chapter. Human nature, although notoriously difficult to delineate, and often easier to define in opposition to that from which it must be defended, relates to some fixed essence shared by (and only by) all humans, "some essential quality that has always underpinned our sense of who we are and where we are going, despite all evident changes that have taken place in the human

---

[5] The suggestion that coercion and autonomy are much more complex notions than is often allowed for in pro-enhancement theories, and that the distinction between the old and the new eugenics is far from clear, is made more convincingly in my opinion by the philosopher and bioethicist Rob Sparrow. In "A Not-so-New Eugenics: Harris and Savulescu on Human Enhancement" (2011), he explores the tension in the works of these theorists between a consequentialist approach – the claim that we should act to increase the amount of welfare in the world (using enhancements) – and their libertarian conclusions – the right of individuals to do as they please. These clash he argues, insofar as ultimately the logic of human enhancement will compel people to enhance their children, and to enhance them according to a standard defined by socially shaped ideals of health and beauty. He writes:

> If parents acted on the obligation that Harris and Savulescu champion, then the result would be a world eerily similar to that dreamed of by previous generations of eugenicists. According to their accounts, in any given society parents should all aim to have the same sort of child, where the nature of this "best baby" is properly sensitive to the prevailing bigotry of the times. Harris and Savulescu's philosophy also implies that right thinking people should engage in social campaigns to influence the reproductive decision-making of other citizens and encourage them to live up to their procreative obligations. (2011: 39).

condition through the course of history (Fukuyama 2002: 101). Human nature, as opposed to that which can be manufactured, chosen or perfected – to the realm of artifice – is grounded in lexicon of givenness, authenticity and continuity.

The notion of givenness is of particular significance. It suggests that an individual life is pre-determined and unpredictable, and that altering, choosing and being able to foresee what life would be is detrimental on many levels. Kass writes:

> Most of the given bestowals of nature have their given species-specified natures: they are each and all of a given sort. Cockroaches and humans are equally bestowed but differently natured. To turn a man into a cockroach – as we don't need Kafka to show us – would be dehumanizing. To try to turn a man into more than a man might be so as well. We need more than generalized appreciation for nature's gifts. We need a particular regard and respect for the special gift that is our own given nature. (2002: 48)

The terminology of givenness is noteworthy here, because as this quote shows, there is often an extension, if not a slippage, of the "given" nature of life in a biological sense, as something that is not chosen, to a more spiritual or religious understanding, imbued with moral undertones, that life, traits and talents are "gifts".

Michael Sandel (2004) has written at length about the given and the gifted nature of being human, arguing for a secular appreciation of the gifted quality of life that enhancement technologies undermine. These technologies represent what he sees as a flawed vision of freedom, the ultimate expression of a drive to mastery that "misses and may even destroy … an appreciation of the gifted character of human powers and achievements" (5). For Sandel, this hubris has the potential to transform three major aspects of human ethics: humility, responsibility and solidarity. First of all, failing to appreciate the gifted nature of life undermines a basic humility that invites humans to "abide the unexpected, to live with dissonance, to rein in the impulse to control" (9). The importance Sandel confers to humility arises from the assumption, frequently advocated by dystopic posthumanists, that there is a "wisdom in nature", and that living creatures are marvels of evolved complexity that human beings will never be able to entirely grasp, a fact that we ignore at our peril. Secondly, the greater control we achieve over our genetic traits, the less we will understand these as contingencies that some individuals were luckily endowed with and others not. We will then come to be seen as increasingly responsible for our talents, as well as our flaws, which will become the burden of each individual rather than of society as a whole. In this view, the unsuccessful will no longer be viewed as disadvantageous, but as responsible of their fate. Thirdly, as our genetic endowments increasingly become a matter of achievement for which we can claim credit, the hard-won value of social solidarity will deteriorate, insofar as it is based in the contingent character of talents and misfortunes that societies as a whole attempt to make up for. This is to say that the more we are aware of the gifted nature of our lot, to the contingencies that enable some individuals to "get ahead", the less willing we will be to share our fate with others.

As the title of Sandel's essay implies, *im*perfection is a constitutive aspect of human existence, and the necessity to deal with it in all of its human manifestations – need, frustration, lack and tragedy, has led to the development of a number of skills and virtues, such as compassion, stamina, courage, irony, humor, and gratitude.

These have, in turn, become valuable in themselves – there is something authentic about imperfection that makes it meaningful in itself. One of the arguments raised against the use of mood-enhancers, in this sense, ensues from the view that feelings of sadness, despair, mourning, etc., are legitimate aspects of our humanity, perhaps even of our evolutionary adaptation (Nesse 1999). They are held to be elements of a complex emotional response that it is normal and appropriate to have towards the world, responses that can only be eliminated at a cost, that of the recognition of finitude and the judgment of one's life in the knowledge that one is going to die. Happiness, in the same vein, is held to be an authentic state of being that must be achieved or strived for, by means of religious or philosophical reflection, a life well-lived, or other meaningful experiences, not via chemical modification. Taking a "happy pill" in this framework would prevent one from dealing with the difficulties and trials that are an integral part of being human. In this sense, the dystopic post-humanist perspective raises concerns that technology will change the world as we know it to such an extent that these values and virtues will lose their conditions of possibility. Of the four types of posthumanism discussed in this book, dystopic posthumanism is the only one that seriously addresses these life-ethical themes. This is perhaps its most important contribution to posthumanist discourse, and a challenge that should be taken up by any critique of it.

For Habermas (2003), like for Sandel, the givenness or the contingency of our genetic makeup acts as a starting point for the self that resides necessarily beyond human will – of the self to the self, or of others to that self. Habermas' position is similar to other dystopic posthumanist writers in that he sees a threat to human nature in enhancement technologies, but it is somewhat more intricate insofar as he is not just concerned with arguing for the sanctity of human nature, but with showing that there is a conspicuous inconsistency between the project of human enhancement and the project of modernity (Fenton 2006). As we have seen, Habermas argues that humans come to share in human dignity when they become members of a moral community, a community of individuals who recognize themselves as equals and accept a set of rules for living together. This moral community is both the proper location and outlet for human nature according to Habermas, it is here that human dignity attains significance, in "interpersonal relations of mutual respect, in the egalitarian dealings among persons" (2003: 33).

This moral community is threatened by genetic enhancement or selection, because it disrupts otherwise equal relationships. First between parents and off-spring: as reproduction becomes an artificial process of choosing what types of children to have, children will see themselves as "products" of parents, whose goals and expectations will have been engineered right into them. For Habermas, a child whose characteristics have been pre-determined before birth, who has been "manu-factured" in such a way, is not free in the way a child whose characteristics are the result of chance is, because she loses the possibility to take a "revisionist stand" towards the "sedimented expectations" of her parents and to choose a life of her own. Thus, if for pro-enhancement advocates it is the freedom of parents (to express their own values and conceptions of the good life) that is at stake, for Habermas, it is the freedom of the child, to choose a life of her own and understand herself as the author of that choice, that should be the focus of our concerns. This is because, secondly, if, aware of having been "manufactured" in line with the expectations of their parents, children will not be able to enter into a relationship of moral equality with their parents, they will not be able to do so with others in the moral community of human beings either.

Thus, by curtailing the freedom of individuals to understand themselves as the undivided authors of their own lives, genetic enhancements threaten the foundation of the moral community as a whole. Habermas' critique of liberal eugenics, then, focuses on bringing to light the inconsistency that he sees at its core: that claiming to protect the autonomy of individuals to realize personal conceptions of the good life, it gives rise to individuals who will be deprived of that autonomy and will no longer be able to participate in the moral community upon which the liberal ideal is predicated. As presented briefly here, this argument against enhancement is more subtle and intricate than other dystopic posthumanist ones insofar as it is not so much based on a sanctity of human nature, but on showing how enhancement technologies are contrary to the very understanding of human nature and the species-ethics that our liberal normative discourse itself assumes. But underlying Habermas' critique is still an argument for human nature (Fenton 2006), as something fixed and given that can be unequivocally distinguished from the realm of the artificial and manufactured, augmented by the normative claim that human nature *should* be distinguished (and so protected) from the realm of the artificial because it is intrinsically valuable.

3.5 Methodological Arguments: The Wisdom of Repugnance Versus Epistemological Strategies

The human enhancement debate is further polarized between dystopic and liberal posthumanists as a dispute about *how* the debate should best proceed, that is, what the best terms for its articulation are. Advocates of enhancement often criticize opponents for grounding their appeal to the special value of human nature in intuitions and emotions, rather than in rational argument. The claim here is that intuitions are subject to various cognitive biases that render them unreliable and prevent them from becoming grounds for rational arguments (Roache and Clarke 2009). This claim builds upon the common theme in dystopic posthumanist discourse of "the wisdom of repugnance", a term first coined by Leon Kass in his 1997 *New Republic* article by the same name, that describes the belief that intuitive negative responses to biotechnological practices are evidence of the intrinsically unethical character of those practices, even if such repugnance cannot be made explicable through reason. Indeed, dystopic posthumanists do not make fully explicit why enhancement is so objectionable, and they often acknowledge the difficulty they have to express in words what is essentially an immediate gut reaction. Sandel writes,

> When science moves faster than moral understanding, as it does today, men and women struggle to articulate their unease. In liberal societies they reach first for the language of

autonomy, fairness and individual rights. But this part of our moral vocabulary is ill equipped to address the hardest questions posed by genetic engineering. The genomic revolution has induced a kind of moral vertigo. (2004: 1)

And Kass: "It is difficult to put this disquiet into words. We are in an area where initial repugnances are hard to translate into moral arguments" (2003: 17). Neither is the notion of human nature or the special human quality or essence that seems to be menaced by enhancement technologies itself ever explicitly spelled out. Fukuyama (2002a), in a rare attempt to do so, introduces the "Factor X", but then merely projects the vagueness of the special quality of human nature onto this term, conceding that it too will always remain somewhat of a mystery:

> When we strip all of a person's contingent and accidental characteristics away, there remains some essential human quality underneath that is worthy of a certain minimal level of respect – call it Factor X. (149)

But further,

> there is no simple answer to the question, What is Factor X? That is, Factor X cannot be reduced to the possession of moral choice, or reason, or language, or sociability, or sentience, or emotions, or consciousness, or any other quality that has been put forth as a ground for human dignity. It is all of these qualities coming together in a human whole that make up Factor X. (171)

Needless to say, for dystopic posthumanists the requirement to engage in an analytic discussion about these issues is itself unacceptable and morally suspicious. Kass insists:

> Can anyone really give an argument fully adequate to the horror which is father-daughter incest (even with consent), or having sex with animals, or mutilating a corpse, or eating human flesh, or even just (just!) raping or murdering another human being? Would anybody's failure to give full rational justification for his or her revulsion at these practices make that revulsion ethically suspect? Not at all. On the contrary, we are suspicious of those who think that they can rationalize away our horror, say, by trying to explain the enormity of incest with arguments only about the genetic risks of inbreeding. (1997: 20)

In this ingenious turnaround, it is rather the requirement to apply methods of human reasoning than the failure to do so that is dubious. The inability to articulate clearly what human nature is becomes evidence that nature and human nature are categories that cannot be reduced to the sphere of ethics governed by human reason. In other words, if it cannot be articulated, than we should not be meddling with it, and the attempt to rationalize this is precisely a symptom of the hubris that leads human beings to believe that they can master and manipulate nature for their own purposes.

Nevertheless, it seems that dystopic posthumanists are often unaware of inconsistencies and faulty reasoning upon which they base their claims. For example, throughout *Our Posthuman Future*, Fukuyama uses evolutionary reasoning when it is convenient and discards it when it conflicts with his conclusion that human beings are special. Fukuyama's claim for human uniqueness also has a clearly tautological nature, according to which (a) humans are unique because they have human nature, in some obvious way distinct from technology, (b) this common human essence is currently under threat by biotechnologies, and (c) in order to preserve human uniqueness, human nature must remain free of technological intervention.[6] Similarly, it is difficult to disentangle the factors of what Sandel presents as a commonsensical if not causal relationship between the drive to mastery over nature and the importance of appreciating the gifted nature of life. Throughout *The Case Against Perfection* he claims both that our Promethean drive prevents us from appreciating the contingent, gifted character of our talents and achievements, and that failing to appreciate the gifted quality of life lets our Promethean drives rein free.

The inconsistencies that ensue when one tries to break down these important dystopic posthumanist arguments proceeds from an even deeper confusion regarding the definition of nature, and slippages back and forth between different definitions. On one use, nature is the world of fossils and fauna, time, climate, plants, all that exists apart from the artificial. We can call this "external" nature. External nature differs from a second definition of nature as "universal": the nature of everything that exists in the universe, including human experience. To add to this, a third kind of nature relates to the nature of things, as an end or telos, the development of a thing, free of external interference. Both external and universal nature then, are seen as having "a" nature in this third sense, and it is mainly this third kind of nature which dystopic posthumanists appeal to.[7] Inconsistencies abound when slippages – mainly from external to universal nature – take place, as they often do in dystopic posthumanist discussions of human nature. Thus the authority of nature as a source of social norms and human conduct derives from its assumed externality to human intervention; but when this kind of external nature is appealed to in the realm of human or social behavior, it necessarily invokes the notion of universal nature, the belief that human and non-human nature are similar enough so that the former is somehow based on or incorporated by the latter. Dystopic posthumanists, (as liberal posthumanists), seem quite oblivious to these slippages, and we shall see how radical posthumanists do a much better job of prying apart all the implications involved in the use of the concept of nature.

For liberal posthumanists, the difficulty that dystopic posthumanists encounter when trying to define what human nature is, and its correlate, the difficulty of articulating what it is that is so repugnant about human enhancement, indicates that the arguments of this approach are intuitive and irrational. In other words, if it cannot be articulated, than it cannot be the grounds of an argument, and should be dismissed. This explains the significant effort undertaken by some pro-enhancement theorists to formulate complex analytical tools and methods for the evaluation of particular technologies. What the complex debate on enhancement technologies requires, they argue, is an analytical heuristic rather than an all-engulfing moral one:

> Whether we should employ a particular enhancement depends on the reasons for and against a particular enhancement ... it is time to take a further step, from asking "Should we do it?" to analyzing the "it" and asking a number of much more specific questions about concrete actions and policy options related to particular enhancement issues within a given

---

[6] Katherine Hayles makes this claim in her article "Computing the Human" (2005).

[7] Noel Castree (2001) discusses these three types of nature at some length.

sociopolitical-cultural context. The result of this will not be a yes or a no to enhancement in general, but a more contextualized and particularized set of ideas and recommendations. (Savulescu and Bostrom 2009: 19)

Intuitive disgust or repugnance cannot, they argue, act as a basis for rejecting such different practices as sports doping, creating superimmunity to biological and viral threats, selecting genetic traits of offspring and taking a pill to improve concentration. Bostrom and Sandberg (2009), for example, develop an evolutionary heuristic they call the "Evolutionary Optimality Challenge", that provides three categories for evaluating an enhancement each which act as limitations on the idea that there is an ungraspable "wisdom of nature". In the same vein, much of the research carried out at the Future of Humanity Institute, the transhumanist think tank led by Bostrom at the University of Oxford, also reveals an emphasis on methodology and epistemological concerns. Here scientific theory and data, techniques of analytical philosophy, statistics, and models of risk thinking, are advanced as the preferred means (and alternative to gut feelings) for examining the questions surrounding enhancement and emerging biotechnologies.[8]

3.6 The Argument Against Human Nature as an Argument for Human Nature: Liberal Posthumanism

The attempt made by liberal posthumanists to develop ways to assess each technology in its own right, in a methodological manner, is valuable and imperative. But it should not act as a means of diverting the discussion away from notions that are difficult to articulate, like nature and human nature, and that are nonetheless being presupposed. A closer look at pro-enhancement arguments quickly reveals that they too appeal to a (normative) conception of nature and human nature.

Dystopic posthumanism often draws a distinction between environmental and genetic transformations in the context of arguments from human nature against enhancement. Fukuyama defines human nature as "the sum of behaviors and characteristics that are typical of the species arising from *genetic* rather than *environmental* factors" (Fukuyama 2002: 130, emphases added). Genetic manipulations, no matter how small, seem to have an impact that cannot be undone in the way that

---

[8] This includes the recognition that an attempt to define what it means to rationalize correctly *before* even approaching the assessment of a new technology, is crucial. One of the institute's four research programs, for example, is called "applied epistemology and rationality". Under this description we find:

How can we become wiser? Answering this question involves looking closely at the way we judge importance and make decisions. It requires close attention to methodology and methodological innovation, particularly ways to improve probabilistic estimation. The shortcomings of extant methodologies is a chief reason why progress on understanding big picture questions for humanity has been slow … Becoming fluent in the language of uncertainty and probability is an important prerequisite for meaningful engagement with many of the problems we work on. (See http://www.fhi.ox.ac.uk/research)

environmental impacts can. In this framework, parents' efforts to influence their children's development, by providing an adequate environment that will be conducive for their development of certain abilities and character traits, is of a completely different category than genetically selecting for or engineering traits and characteristics. The former is something that we *expect* of parents, as part of a child's education and the provision of opportunities, while it is the latter is seen as morally unacceptable. For Habermas, who attempts to ground this distinction in perhaps the most rigorous manner, "natural fate" and "socialization" differ in a moral sense because we have a "fundamentally different kind of freedom" towards either (2003: 14). This is because he believes that we can reject or at the least revise our socialization (with psychotherapy for example) in ways that we cannot reject or revise parental expectations in the form of genetic manipulations that have been carried out before our birth. As opposed to what parents do when they shape a child's environment, genetic interventions "have the peculiar status of a one-sided and unchallengeable expectation" (51).[9]

For liberal posthumanists, drawing on scientific research on the developmental roles of genes and environment, and the complex interplay between them, this distinction is seen as highly problematic. Environmental influences like education or nutrition are seen as no "softer" than genetic influences, so that the attempt to improve people by modifying their environment or by modifying their genes is morally equivalent (Agar 2004: 172). In this sense, genetic interventions can be likened to other types of child-rearing enhancement techniques: tutors, camps, training programs, special diets, etc. Harris writes:

> if the goal of enhanced intelligence, increased powers and capacities, and better health is something that we might strive to produce through education ... why should we not produce these goals, if we can do so safely, through enhancement technologies or procedures? If these are legitimate aims of education, could they be illegitimate aims of medical or life science? (2007: 2)

What's more, neither types of influence are seen as *determining* a child's future. Agar writes, "Although genomic information may give parents the power to influence the probability that a given life plan will be chosen, it is unlikely that the probability could ever be raised to the point of reliability" (125). Most importantly, if the claim for a moral parity of genetic and environmental engineering can be upheld, and these are really just two different means of "manufacturing" humans, then it becomes easy to claim for parity on other levels – such as that we have, as humans, *always* been in the business of human modification and engineering. It is in this sense that PGD is often compared to a more technological or precise form of the mate selection that humans undertake when they seek sexual and/or romantic relationships, consciously or unconsciously assessing the genetic qualities of their partner. This kind of argument from precedent, what Erik Parens (1995) has called the

---

[9]The charge of genetic determinism is often raised here. Why would we be free to revise our socialization but not our genetic enhancement? And although Habermas is well aware of this objection, his arguments against it are not very clear, focusing on the "intention" governing the genetic intervention (124, n. 54).

"we've already done it (and everything's been okay) argument", holds that if practice X has been morally acceptable in the past, and if practice Y is just like practice X, then practice Y should be morally acceptable now and in the future. Here it is the continuities, not the differences, between old and new methods of enhancement that are emphasized. Hence, in addition to comparing PGD to dating, taking Ritalin for enhanced cognitive performance can be compared to drinking a strong cup of coffee or to getting a good night's sleep, and taking Prozac can be compared to older forms of mood manipulation like religious rituals.

Crucial to this line of reasoning is a rejection of the idea of an intrinsic goodness of nature, or of what nature has "given". Put bluntly, what case can be made for species-specific capacities like rape, genocide, torture and racism? Bostrom, in response to Fukuyama's "factor X" definition of the human essence, writes,

> There is too much that is thoroughly unrespectable in human nature (along with much that is admirable), for the mere fact that X is a part of human nature to constitute any reason, even prima facie reason, for supposing that X is good. (2009: 126)

There is no obvious reason why that which is "given" at conception is in some way more real, true or moral and should be left untouched in this view. For liberal post-humanists, human nature as such has no moral authority. Furthermore if genes are distributed through the random workings of natural selection, it is difficult to argue that anyone "deserves" the genes they were born with, or that those born with "bad" genes are undeserving in any way, and *ought* to "suffer" them. If chance rather than merit is the decisive factor here, than morality may well be on the side of those who would want to help those who lucked out in life's genetic lottery: "In the face of these staggering odds of an indifferent universe," Steven Pinker has claimed, "I would suggest that anything that gives us a leg up in this struggle, that increases our odds by some increment, should be welcomed" (2004).

The arbitrariness of evolutionary mechanisms thus acts as a justification – if not a plea – for intervention and enhancement, and anyone who fails to see this is committing not only a naturalist fallacy, but a pre-Darwinian one. This is the charge made in some detail by Allen Buchanan (2011b) for example. Buchanan compares the anti-enhancement stance in this context to a deluded view of evolution as a "master engineer", where "organisms are like engineering masterpieces: beautifully designed, harmonious, finished products that are stable and durable (if we leave them alone)" (29). Humans in this view have reached some particularly valuable end-point, an evolutionary summit of perfection that should not be tampered with. But this "rosy pre-Darwinian" view of evolution is misguided, Buchanan argues. Rather, evolution produces sub-optimal designs, it is largely insensitive to post-reproductive quality of life, it is driven in large part by drift and local optimization and it "selects for fitness, not human good" (48). Evolution, as Darwin theorized, is more like a "grim tinkerer", and organisms – including humans in their current evolutionary state – are products of random mutations and selection, "cobbled-together, unstable works in progress" (28). It is quite absurd, then, to take the result of this fickle and unintended evolutionary process as it stands so far as some ideal to be defended. And Buchanan's aim is not only to provide reasons to reject the "master engineer" analogy in favor of the "grim tinkerer" analogy, but to use this as grounds for arguing that it may be preferable, at least in some circumstances, to actively pursue genetic enhancements rather than leaving the development of the human species entirely to the clumsy, wasteful and often flawed workings of evolution:

> We have to steadfastly resist the common tendency to think that the latest product of the evolutionary process is the best, either biologically speaking, or in terms of human values. We can't say we are the best in either sense, and that's why we should take the prospect of biomedical enhancement seriously. (47–48)

The appeal to humility, it seems, is now on the side of the advocates of enhancement.

This type of argument from nature is a means of invalidating essentialist appeals to human nature as something that is fixed and should be preserved as it currently stands; it dismisses the anti-enhancement approach as a call to safeguard the "status quo" just because it is the status quo. But it also easily translates into a prescriptive argument for human nature in itself, by which (a) the human species is constantly changing and improvable, (b) that its very existence implies affecting its surrounding and itself in unpredictable ways and (c) that the aspiration to self-improvement is an integral part of this dynamic. To "choose to be better", as Savulescu writes, "is to be human" (Savulescu et al. 2004: 670). This is troubling, insofar as liberal posthumanists are usually so adamant in their critique of appeals to human nature as the basis of normative arguments against enhancement, that it is the last place one might expect to run into arguments from human nature. What's more, when liberal posthumanists do appeal to a conception of human nature, tacitly or openly, it is a conception that is just as normative as the ones they attack (Hauskeller 2009).

Gregory Stock (2002), for example, talks about our Promethean nature that compels us to continue "stealing fire from the Gods". Shunning our ability to manipulate gene pools and to engineer germ-lines, he claims, "would be to deny our essential nature and perhaps our destiny ... such a retreat might deaden the human spirit of exploration, taming and diminishing us" (2002: 170). This normative and essentialist understanding of the human is common among liberal posthumanists. In this view humans have always manipulated their environments and designed tools to increase their chances of survival, from shoes and clothing to numeracy and literacy, enhancement is a vital part of what we do. Savulescu and Bostrom write:

> *all* learning could viewed as physiological enhancement, and *all* physical and organizational capital could be viewed as external enhancements. Stripped of all such "enhancements" it would be impossible for us to survive, and maybe we would not even be fully human in the few short days before we perished. (2008: 3)

Radical self-transformation, in this view, is intrinsic to our human nature, and to ban the use of available means to achieve it is to contradict that nature, a "betrayal of the dynamic inherent in life and consciousness" (More 1996: 8). Human nature in these accounts functions in the realm of potential, what it is to be human is about the strive towards a future realization of a better human, and enhancements, whether they be old or new, environmental or technological, are instrumental for turning humans into what they are meant to be. They make us more, rather than less, human. As Haukeller explains, on this very clear view of human nature in pro-enhancement discourse,

The potential to become something other than what we are is thus not only what makes us human, but also what gives us that special worth on which all our moral rights ultimately depend. Hence, to turn our backs on this potential would both violate our nature and compromise our dignity. (2011: 47)

Thus, while liberal posthumanists vehemently oppose the appeal made by dystopic posthumanists to human nature and its moral relevance as reference to their objection to enhancement technologies, it is clear that they too make normative presuppositions about human nature and the need to defend it. Both camps anticipate an injury to human nature as a result of the consent to or ban on enhancement technologies, in which humans will become "less than human" on the one hand, or "not fully human", as Savulescu and Bostrom write, on the other.

3.7   Conclusion

The intensity of the human enhancement debate becomes more understandable when it is framed in terms of a profound and incommensurable disagreement about the nature of human nature, rather than "merely" about issues of safety, access or technical feasibility. Furthermore, framing the debate in these terms is important because, as argued in the previous chapter, while the explicit and implicit accounts of human nature presupposed by dystopic and liberal posthumanism seem to be at odds, they are really two versions of modern liberal humanism and its ontological dualism, by which humans, by virtue of some human essence, have a unique status that separates them from the rest of the natural (and non-natural) world. Dystopic and liberal posthumanists ascribe to a fairly similar model of the human, as a being that is essentially autonomous from its environment and its technologies. It is this understanding of the human as an independent, autonomous entity with clear boundaries that underlies both dystopic and liberal posthumanism's objection and espousal of enhancement technologies. For dystopic posthumanists technology is seen as impinging on the human from an outside: technologies that do not respect the boundaries of the subject and penetrate its autonomous constitution are conceived as a potential threat to human freedom, individuality and dignity. For liberal posthumanists the human uses technology to master that outside: here too the subject is an independent, autonomous entity whose nature may be dynamic to a point, insofar as it continuously integrates new technologies into its experience and constantly aspires to self-improve, but is simultaneously fixed in a transcendent position vis-à-vis its environment. In order to move beyond the impasse that the enhancement debate comes up against what is needed is a non-humanist approach.

References

Agar, N. (2004). *Liberal eugenics: In defence of human enhancement.* Oxford: Blackwell.
Bostrom, N. (2002). Existential risks: Analyzing human extinction scenarios and related hazards. *Journal of Evolution and Technology,* 9(1). Available at http://jetpress.org/volume8/symbionics. html. Accessed 20 August 2013.

Bostrom, N. (2003). The transhumanist FAQ, Version 2.1. http://www.transhumanism.org/resources/FAQv21.pdf. Accessed 6 June 2013.

Bostrom, N. (2009). Why I want to be a posthuman when I grow up. In B. Gordijn & R. Chadwick (Eds.), *Medical enhancement and posthumanity* (pp. 107–137). Heidelberg: Springer.

Bostrom, N., & Sandberg, A. (2009). The wisdom of nature: An evolutionary heuristic for human enhancement. In J. Savulescu & N. Bostrom (Eds.), *Human enhancement* (pp. 375–416). Oxford: Oxford University Press.

Buchanan, A. (2011). *Beyond humanity? The ethics of biomedical enhancement*. Oxford: Oxford University Press.

Buchanan, A., Brock, D. W., Daniels, N., & Winkler, D. (2000). *From chance to choice: Genetics and justice*. Cambridge: Cambridge University Press.

Castree, N. (2001). Socializing nature: Theory, practice and politics. In N. Castree & B. Braun (Eds.), *Social nature: Theory, practice and politics* (pp. 1–21). Oxford: Blackwell.

Elliot, C. (2003). *Better than well: American medicine meets the American dream*. New York: W.W. Norton & Company.

Fenton, E. (2006). Liberal eugenics & human nature: Against Habermas. *Hastings Center Report, 36*(6), 35–42.

Franklin, S., & Roberts, C. (2006). *Born and made: An ethnography of preimplantation genetic diagnosis*. Princeton: Princeton University Press.

Fukuyama, F. (2002). *Our posthuman future: Consequences of the biotechnology revolution*. New York: Farrar, Straus and Giroux.

Habermas, J. (2003). *The future of human nature*. Cambridge: Polity.

Harris, J. (1993). Is gene therapy a form of eugenics? *Bioethics, 7*, 178–187.

Harris, J. (2007). *Enhancing evolution: The ethical case for making better people*. Princeton: Princeton University Press.

Hauskeller, M. (2009). Prometheus unbound: Transhumanist arguments from (human) nature. *Ethical Perspectives, 16*(1), 3–20.

Hauskeller, M. (2011). Pro-enhancement essentialism. *AJOB Neuroscience, 2*(2), 45–47.

Hayles, N. K. (2005). Computing the human. *Theory, Culture and Society, 22*(1), 131–151.

Hubbard, R., & Newman, S. (2002). Yuppie genetics, ZMagazine, March. http://www.zcommunications.org/yuppie-eugenics-by-ruth-hubbard-and-stuart-newman.html. Accessed 21 Sept 2013.

Hughes, J. J. (2002). The politics of transhumanism. Available at http://www.changesurfer.com/Acad/TranshumPolitics.htm. Accessed 6 June 2013.

Hughes, J. J. (2009). Social pressures for technological mood management. *Free Inquiry, 29*(5), 28–32.

Kass, L. (1997). The wisdom of repugnance. *The New Republic, 216*(22), 17–26.

Kass, L. (2002). *Life, liberty, and defense of dignity: The challenge for bioethics*. San Francisco: Encounter Books.

Kass, L. (2003). Ageless bodies, happy souls: Biotechnology and the pursuit of perfection. *The New Atlantis, 1*(Spring), 9–28.

Kass, L., & The President's Council on Bioethics. (2003). *Beyond therapy: Biotechnology and the pursuit of happiness*. New York: Regan Books.

Kramer, P. (1993). *Listening to Prozac: A pysychiatrist explores antidepressant drugs and the remaking of the self*. New York: Penguin.

Lykken, D. (1999). *Happiness: The nature and nurture of joy and contentment*. New York: St. Martin's Press.

Lykken, D., & Tellegen, A. (1996). Happiness is a stochastic phenomenon. *Psychological Science, 7*(3), 186–189.

Miah, A. (2009). Make me a superhero: The pleasures and pitfalls of body enhancement. *The Guardian* (May 1). http://www.guardian.co.uk/science/2009/may/01/body-enhancement-cosmetic-surgery-genetics. Accessed 6 June 2013.

More, M. (1996). Transhumanism. Towards a Futurist philosophy. http://www.maxmore.com/transhum.htm. Accessed 6 Jun 2013.

Nesse, R. (1999). Is depression an adaptation? *Archives of General Psychiatry, 57*, 14–20.

Parens, E. (1995). Should we hold the (germ) line? *Journal of Law Medicine and Ethics, 23*(2), 173–176.

Pinker, S., Sandel, M., Coffin, B., & Glickman, D. (2004). *The new eugenics? The brave new world of designer children, bionic athletes, and genetic engineering.* Boston: Harvard University Institute of Politics. Available at "http://forum.iop.harvard.edu/content/new-eugenics-brave-new-world-designer-children-bionic-athletes-and-genetic-engineering". Accessed 6 Jun 2013.

Roache, R., & Clarke, S. (2009). Bioconservatism, bioliberalism, and repugnance. *Monash Bioethics Review, 28*(1), 1–21.

Sandel, M. (2004). *The case against perfection* (pp. 1–11). April: The Atlantic.

Savulescu, J. (2001). Procreative beneficience: Why we should select the best children. *Bioethics, 15*(5), 413–426.

Savulescu, J. (2005). New breeds of humans: The moral obligation to enhance. *Ethics, Law and Moral Philosophy of Reproductive Biomedicine, 1*(1), 36–39.

Savulescu, J. (2007). In defence of procreative beneficence. *Journal of Medical Ethics, 33*(5), 284–288.

Savulescu, J. (2008). The perils of cognitive enhancement and the urgent imperative to enhance the moral character of humanity. *Journal of Applied Philosophy, 25*(3), 162–167.

Savulescu, J. (2012). The maverick: 'It's our duty to have designer babies'. *Reader's digest* (September). http://www.readersdigest.co.uk/magazine/readers-digest-main/the-maverick-its-our-duty-to-have-designer-babies. Accessed 6 June 2013.

Savulescu, J., & Bostrom, N. (Eds.). (2009). *Human enhancement.* Oxford: Oxford University Press.

Savulescu, J., Foddy, B., & Clayton, M. (2004). Why we should allow performance enhancing drugs in sport. *British Journal of Sports Medicine, 38*(6), 666–670.

Sparrow, R. (2011). A not-so-new eugenics: Harris and Savulescu on human enhancement. *Hastings Center Report, 41*(1), 32–42.

Stock, G. (2002). *Redesigning humans: Choosing our children's genes.* London: Profile.

Walker, M. A. (2007). Happy-people-pills and prosocial behaviour. *Philosophica, 79*(1), 93–111.

Walker, M. A. (2009). The case for happy-people pills. *Free Inquiry, 29*(5), 33–36.

Chapter 4
Towards a Non-Humanist Posthumanism: The Originary Prostheticity of Radical and Methodological Posthumanism

Abstract We can now begin to take a closer look at radical and methodological posthumanism as the main candidates for a non-humanist alternative to dystopic and liberal posthumanism. These approaches develop alternative frameworks that move beyond the essentialism inherent in instrumental and substantive models of technology that inform dystopic and liberal posthumanism. Radical posthumanism argues for a reflexive model of technology, in which technologies are both seen as the product of human creativity and a force that shapes human existence, i.e. technologies are determinative of human experience, though not deterministic. And methodological posthumanism introduces the key concept of technological mediation, which implies that technologies are active mediators of how humans experience the world and how humans act, transforming ourselves and the world in the process.

Both approaches imply an "originary prostheticity", the idea that the human exists in relation to and is dependent on its technologies; that the human emerges as a result of this relationship. In this view, the dualist humanist paradigm is a hindrance to understanding how humans engage with technologies. Both approaches also argue for more positive conceptualizations of technology than previous critical philosophy of technology allowed for. For radical posthumanism, starting with the "Cyborg Manifesto", this implies a celebration of the political potential inherent in new technologies. For methodological posthumanism this means conceptualizing the ambivalent status of technology, which may lead to a loss of involvement of humans in their environment in some instances, but also amplifies and creates new forms of engagement.

Keywords Technological mediation • Originary prostheticity • "Empirical turn" • Instrumental model of technology • French philosophical materialism

T. Sharon, *Human Nature in an Age of Biotechnology: The Case for Mediated Posthumanism*, Philosophy of Engineering and Technology 14, DOI 10.1007/978-94-007-7554-1_4, © Springer Science+Business Media Dordrecht 2014

We can now begin to take a closer look at radical and methodological posthumanism as the main candidates for a non-humanist alternative to dystopic and liberal posthumanism. This chapter will present various approaches to technology with a focus on how novel radical and methodological posthumanism's engagement with technologies is. First of all, these approaches reject the overall pessimistic and transcendentalist view of technology as a dehumanizing and alienating force that characterizes classical philosophy of technology. Classical philosophy of technology can be seen as offering essentialist critiques of technology that refer back to foundational narratives of the organic human, an uncontaminated nature or an authentic reality, that echo the dualist paradigm of humanism. These critiques are viewed as nostalgic and escapist by radical and methodological posthumanism, a result of a search for the conditions of possibility of technology as a monolithic phenomenon. This essentialism, I shall argue, not only runs through classical and more contemporary techno-skeptic approaches to technology like dystopic posthumanism, it also underlies technophilic approaches like liberal posthumanism. This becomes clear when we see that the common distinction between instrumental models of technology, in which technologies are seen as neutral tools, and substantivist models of technology, in which technologies have a deterministic transformative effect on humans, is also based in an essentialist understanding of technology as strictly separate from humans.

Instead, both radical and methodological posthumanism develop models of technology that are non-essentialist and allow for the possibility of positive appraisals of technology, thus marking an important turning point in critical philosophy of technology. Radical posthumanism argues for a reflexive model of technology, in which technologies are both seen as the product of human creativity and a force that shapes human existence, i.e. technologies are determinative of human experience, though not deterministic. For radical posthumanists, Donna Haraway's "Cyborg Manifesto" (1991) already sets the tone for a positive, even celebratory, view of contemporary technologies as strategies of resistance against the foundational narratives of modernity. In methodological posthumanism, a non-essentialist model of technology ensues first from an "empirical turn" in which research into the development and use of specific and concrete technological artifacts can account better for the many ways in which human contexts and values shape the use of technology. Secondly, methodological posthumanism, namely through the work of Don Ihde and Bruno Latour, introduces the key concept of technological mediation, which implies that technologies are active mediators of how humans experience the world and how humans act, transforming ourselves and the world in the process. Mediation can thus replace alienation as the central concept for analyzing technology, and lead to a more nuanced view of technology according to which technology offers a form of engagement with the world.

Finally, this chapter explores the more philosophical implications of radical and methodological posthumanism's non-essentialist approaches to technology through the notion of prostheticity. I will argue for a fundamental though subtle distinction between "supplementary" prostheticity, in which technology is seen as an appendage, something that is "added on" to the human all the while leaving the two categories of human and technology largely intact, and "originary" prostheticity as defined by Bernard Stiegler (1998), in which the human is seen as originally existing in relation to and as dependent on its technologies, which it always already incorporates. The discussion on prostheticity will be an occasion to review some other important historical and contemporary theorists of technology, including Ernst Kapp, Maurice Merleau-Ponty, Kevin Warwick, Stelarc and Gilles Deleuze and Félix Guattari. Originary prostheticity will act as a final term through which to grasp the implications of a non-humanist approach to technology, as developed by radical and methodological posthumanism.

The various approaches to technology are positioned in the following chart:

Technology as:	Essentialist (supplemental prostheticity)	Non-essentialist (originary prostheticity)
Neutral	**Instrumentalist** (human mastery over technology)	X
Value-laden	**Substantivist** (mastery of technology over humans)	**Reflexive & mediating** (technologies mediate reality and human behavior)

4.1 Essentialism: Techno-Skeptic and Technophilic, Instrumental and Substantive Approaches

4.1.1 The Alienation Thesis

The philosophical attention given to technology is a relatively new phenomenon, and it is only quite recently, no earlier than the second part of the twentieth century, that the philosophy of technology can really be identified as a discipline in itself (Achterhuis 2001; Mitcham 1994). In the twentieth century, an interest in technology began taking shape mainly in the inter-war period, in the works of theorists like Frederich Dessauer (1927), Karl Jaspers (1933) and Ortega y Gasset (1939) in Europe, and John Dewey (1929) and Lewis Mumford (1934) in the United States. For this first generation of philosophers of technology, the general appraisal of technology tended to be rather ambiguous, although most influential European theorists advanced a rather pessimistic appraisal of technology, inspired by the novel dimensions of industrial growth exemplified in factories, assembly lines, Taylorism and the horrors caused by the new weaponry of mass destruction introduced in World War I. Writing after World War II, a new generation of critical philosophers of technology, many associated with the Frankfurt School, developed this skeptical and negative assessment of industrial technology and its growing association with capitalism. This new generation, which includes theorists like Jacques Ellul (1965), the later Martin Heidegger (1977a), Hans Jonas (1979), and disciples of the Frankfurt School (Habermas 1968; Marcuse 1964), can be seen as making up the "classical" approach in the philosophy of technology.

The works of these theorists are characterized by a dominantly pessimistic view of technology as a destructive force, and a bleak portrayal of technological culture as leading to processes of objectification and dehumanization. Technology, and the rationalist, instrumentalist way of thinking that underlies it, these theorists argued, encourages individuals to approach reality as raw material rather than something inherently valuable. Many of these theorists feared that human beings, in this leveling of all things, would loose their unique individuality in a mass culture of conformity. In these approaches, technology is seen as standing in opposition to the human and to nature, as something that needs to be controlled or countered by a culture that should recover its connections with its own inner truth and authentic values. At the same time, these early works tended to approach technology in a general, or a transcendentalist way, in which Technology with a big "T" is seen as a monolithic phenomenon. It is this kind of technology that is implied in the ideas of "autonomous" or "runaway" technology, technologies that have gone out of control. In this sense, these theorists were less interested in the materiality of technology, than in the conditions of possibility of technology, of the technological way of "disclosing" things that was implied by the development and use of technologies.

Heidegger's "The Question Concerning Technology" (1977a) is usually taken as the key text in this tradition. In this essay, Heidegger famously argues that technology is a disclosing of reality, and that the essence of technology is a stance towards the world, a "mode of revealing", since humanity brings itself forth in part through its way of using things. Unlike the bringing-forth of classical *poièsis*, for Heidegger modern technology is a "challenging-forth" that enframes nature as a "standing-reserve", as "a coherence of forces calculable in advance" (1977: 303), raw material waiting to be ordered and appropriated. The danger in this form of revealing for Heidegger is that it transforms everything, not only nature but human labor too, indeed, humanity, into standing reserve, all the while concealing or destroying more fundamental ways of revealing the essence of being.[1] Similar themes are shared in many respects by Ellul (1965), for whom technicized society causes a loss of human autonomy in the service of machines; and by Marcuse (1964; 1998), who wrote of the "technological attitude" which transforms the function of individuality into a conformism that is destructive of individuality.

Despite some important differences, namely in political inclination, dystopic posthumanists can be seen as heirs of this critical type of classical philosophy of technology. This is because while theorists like Ellul and Marcuse differ from dystopic posthumanists insofar as their critical take on technology emerges from a

---

[1] Heidegger also sees a "saving power" in technology, which he develops to a much lesser extent than the danger here involved towards the end of the essay. If the human relationship to technology is the result of a challenging-forth that humanity takes up, he argues, then technology is not just a revealing and an ordering that humans orchestrate, it is something that humans have not made, but receive. Revealing is something that does not arise from human ingeniousness, but something that "awaits" humans, that needs them. Humans make revelations not as a means of conquering the world but to show how they belong to the world, even in their apparent alienation from it. Technology's ability to reveal beings and create a world, a capacity that it shares with artwork, can thus also *save* us.

critique of the effect of the rationalization of the world beyond merely technological realms (on democracy, individualism, the private realm, traditional forms of community, etc.), both classical philosophers of technology and dystopic posthumanism imply an *essentialist* critique of technology. In essentialist models of technology, both humans and technology are pitted against each other as essences, where the human is understood as occupying a unique position and possessing an intrinsic nature that differs fundamentally from technology. Essentialism here is thus also another expression of the dualist paradigm of humanism. In classical philosophy of technology, this essentialism is manifested as variations on the "thesis of alienation" (Verbeek 2002), the claim that the growing ubiquitousness of technology and the technological way of thought will alienate humans from what they "really" are or what reality "really" is. Such essentialist critiques refer back to foundational narratives of the organic human or an uncontaminated nature and call for the need to defend a unique human nature from technological intervention.

4.1.2 Instrumental Versus Substantive Models of Technology

It might be tempting to categorize all such techno-skeptic approaches to technology as essentialist, and as we shall see shortly, the more optimistic accounts of technologies suggested by radical and methodological posthumanism proceed precisely from their anti-essentialist positions. But understanding the various contemporary approaches to technology is more complicated than this, because technophilic approaches like liberal posthumanism are also fundamentally essentialist. It is helpful to introduce a more useful distinction here, between instrumental and substantive (or constitutive) models of technology, that was first introduced by the philosopher of technology Albert Borgmann (1984). In this distinction, the instrumental model views technology as a tool or instrument that is used to satisfy needs, while the substantive model attributes meaningful values to technology that make it an autonomous and constitutive cultural force. This differentiation is extremely consequential, both for the philosophy of technology and for social policy that might legalize, normalize and regulate the use of technologies.

   The instrumental model of technology is based on the idea that technologies are mere tools applied to nature, means to ends that have no inherent value in themselves, so that means and ends are independent of each other. If any value is to be attributed to technology, it is only the formal value of efficiency. Here technology is an indifferent, "rational entity" and is universally applicable, thus allowing similar norms of measure to be applied in diverse situations. Any concern about technology in this context relates to its range of efficiency and the danger of its "falling into the wrong hands" (a scenario that is possible precisely because means and ends are independent of each other). In contrast, the substantive model of technology views technologies as much more than value-neutral tools or objects, and attributes values to specific technologies as well as to technology as a whole. Technology is not simply instrumental to various values, it already embodies certain values. In this sense the use of technology for a particular purpose in itself assumes a value choice rather than a merely more efficient means of realizing a pre-existing value. The substantive model implies that technologies have a transformative nature: they are constitutive of human existence. Means here are not independent of ends, but actually form a framework for a way of life that includes ends.

The example of plastic surgery can illustrate how both theories engage differently with technology. From an instrumentalist perspective, plastic surgery is an efficient means which is independent of the ends brought to it by the user, whether this end be "reparative" (following an accident, for example), "enhancive" (in an attempt to boost low self-esteem for example), or "cosmetic" (driven by pure vanity). These problematic differentiations themselves lie beyond the scope of the technology, which has no "preference" as to which uses it can be put to. From a substantivist perspective however, the very fact that the technology of plastic surgery exists creates a social world quite different from the social world in which it did not, one in which problematic differentiations and multiple options are brought into existence alongside the technology itself. Andrew Feenberg (1991) uses the phenomenon of fast food as a substitution for traditional family meals to illustrate how the "value" of efficiency is seen by substantive theory as deteriorating and replacing all other values. The instrumentalist, explains Feenberg, will analyze fast food as an efficient solution to the technical operation of eating, disregarding the cultural implications of this technology, such as the ritual aspect of food consumption. The substantive theorist, on the other hand, will view the decline of the traditional family dinner as one of the unintended cultural consequences of fast food technology.

In Marx's materialist history, for example, insofar as technology transforms the relation between the laboring individual and the method of labor, it is substantive. As a socially structuring force that forms the laboring body in industrial capitalism, it affects consciousness. Marx differentiates between "simple" tools, the kind that we can hold in our hands, and larger machinery and systems of machinery, found in factories. This is a qualitative difference that has more to do with the kind of effect such technology has on the laborer than with a difference in size. While the individual maintains an independent capacity to labor with the former, the latter have a transformative power on the laborer, namely the power of alienation:

> The worker's activity … is determined and regulated on all sides by the movement of machinery, and not the opposite … The science which compels the inanimate limbs of the machinery, by their construction, to act purposively, as an automaton … acts upon [the worker] through the machine as an alien power, as the power of the machine itself. (Marx 1993: 693)

The classical philosophers of technology mentioned above all assume a substantive model of technology too. Heidegger's analysis in "The Question Concerning Technology", for example, proceeds from what can be understood as a substantive critique of the instrumental logic of modern technology. Technology, Heidegger maintains, is a mode of revealing (1977a: 294). For the Greeks, he explains, the mode of revealing (*alètheiai* or truth) was *poièsis*, a bringing-forth of things from concealment to unconcealment, their unfolding according to the four causes or modes of occasioning as defined by Aristotle (*materialis, formalis, finalis, efficiens*). *Technè*, the skill or activity of this bringing-forth, needs to be understood in this sense of revealing, of unconcealing, rather than as manufacturing, making or manipulating, which belongs with *poièsis*. The modern conception of technology, the instrumental, neutral, or what Heidegger also calls the "anthropological" definition of technology, is, in contrast, something entirely new, because its mode of revealing is not a bringing-forth in the sense of *poièsis* but a "challenging" of nature to supply energy to be extracted and stored, an "enframing" which reveals things as "standing-reserve", ready to be ordered, transformed and used. Modern technology is also a mode of revealing, of unconcealment, but one that, by enframing, reveals by reduction to an orderability as standing-reserve, so that, as Heidegger illustrates, the earth is now revealed as that which yields coal or ore, and the River Rhine as that which yields hydraulic pressure that can be transformed, stocked up and distributed. In other words, modern technology shares the capacity for revelation with *technè*, but puts this capacity to different ends.

The danger inherent in this mode of revealing is twofold for Heidegger. First of all, in the mode of enframing, man comes to see himself as the master of nature as standing-reserve and of technology as that activity that turns standing-reserve into energy. But, Heidegger argues, unconcealment is "neither only a human activity nor a means within such activity" (1977a: 302), and the idea that man has control over unconcealment is an illusion of the modern age, rendering the very notion of technology as "instrumental" untenable. Rather, within the mode of revealing of modern technology, man, just as the energies of nature, is also challenged, ordered, and transformed into standing-reserve. Heidegger uses the example of the forester, who appears to "walk the forest path in the same way his grandfather did", but is today ordered by the commercial wood industry, and beyond that the paper industry, and beyond that the written press, whether he knows this or not. Heidegger writes:

> As soon as what is unconcealed no longer concerns man even as object, but exclusively as standing-reserve, and man in the midst of objectlessness is nothing but the orderer of the standing-reserve, then he comes to the very brink of a precipitous fall, that is, he comes to the point where he himself will have to be taken as standing-reserve. Meanwhile, man, precisely as the one so threatened, exalts himself to the posture of lord of the earth. In this way the illusion comes to prevail that everything man encounters exists only insofar as it is his construct. (1977a: 308)

Enframing does not only pose this danger of all beings including man becoming standing-reserve, it furthermore obscures any other possible ways of revealing, ones that might be more original and more truthful, namely *poièsis*. This is a threat to what Heidegger sees as the freedom of mankind – the free relationship mankind has to itself in light of the recognition that it brings itself forth in order to be – and the loss of what is most essentially human, the capacity for new revealings. This is, according to Heidegger, the danger "in the highest sense" posed by the instrumental view of technology: "enframing ... threatens to sweep man away into ordering as the supposed single way of revealing, and so thrusts man into the danger of the sur-render of his free essence" (1977a: 313–314).

In the substantive model, the instrumental logic of technology is always more than just an attitude towards technology. It is usually perceived as a new type of cultural system, one that turns the social world into an object to be dominated. For Ellul (1965), for example, this instrumental logic is relentlessly expanding, absorbing every pre-technological form of social life, and becoming the defining characteristic of society. The artificial realm of *technique*, in this framework, replaces nature as the new milieu of contemporary society. Traditional values cannot survive the challenge of modern technology and once the path of technological development is taken, societies are inevitably transformed into "technological" societies. This also supposes that technology develops autonomously, that it is a self-governing force unto itself, that once unleashed, threatens to take over all domains of social life. This view of genetic engineering technology as an autonomous force is repeated in many of the dire predictions of dystopic posthumanists. Here biotechnology is attributed its own agency. No longer under our control, it has the potential to undermine our social relations, to disrupt our political and legal norms, and to change our very nature – and this at a pace so fast that we may not even recognize it. The biotech revolution is out of control, moving ahead too quickly and without our consent.

The instrumental and the substantive models of technology differ greatly, and as is clear from these examples, substantive theorists often develop their models in line with a critique of what is seen as a very detrimental instrumental understanding of technology. But both the instrumental and substantive models imply a predominantly essentialist understanding of technology, in which humans and technologies have an underlying, unchanging essence. In the instrumental view, the idea of technology as an instrument implies hermetic boundaries between a human self and a tool that it puts to use. The substantive model, though more complex because of its emphasis on value-ladenness and the transformative or constitutive function attributed to technology, is also informed by an essentialist critique of technology, in which nature and humans stand in opposition to technology, and share an authenticity and organic pureness that is threatened by technology and the instrumental mode of reasoning that it promotes. Essentialism in these terms is another way of expressing the humanist dualist paradigm, with its consequent philosophical implications for understandings of human nature. It also implies, for both instrumental and substantive models, what Andrew Feenberg (1991: 8) calls a "take it or leave it" attitude towards technology that has significant implications for technology design and regulation, and public debate about technologies in general. This is because in the instrumental model, as Feenberg explains, only the range and efficiency of a technology's application is subject to debate, insofar as technology remains indifferent to values. While in the substantive model, technology enjoys an autonomy that implies a determinism that leaves us quite helpless once it is let loose. In order to prevent such "technology is destiny" attitudes, non-essentialist models should be explored.

4.2   Overcoming Essentialism and Doing Away with Pessimism: Reflexive Technology and Technological Mediation

Both radical and methodological posthumanism reject the essentialist assumptions of instrumental and substantive models of technology. They also mark a turning point in critical theory from a generally negative to a more positive assessment of technologies that proceeds from their anti-essentialist position – and so differentiates them from simple technophilic instrumentalist approaches like liberal posthumanism. These approaches introduce important concepts for the analysis of technology in the form of *reflexivity* and *technological mediation* that allow them to move beyond essentialism, instrumentalism and determinism and to forge alternative means of conceptualizing new technologies and human-technology relationships.

4.2.1   The Cyborg Versus Organicism

For radical posthumanism, an optimistic tone is already set in Donna Haraway's "Cyborg Manifesto" (1991). Located in its historical context of the mid-1980s, the Manifesto denotes a rift in feminist theory's conceptualization of nature, and attempts to challenge "organicist" feminist approaches to nature. To understand this rift and the significance it bears for radical posthumanism's approach to emerging biotechnologies, it is important to understand the Manifesto as part of the revisionist project that had been undertaken by feminist critiques of science. Feminist scholars of science have argued that biological science is not an empirical and objective account of the world, but a form of constructed knowledge that is intimately tied to power effects (Keller 1985; 1995; Harding 1986, 1991; Hubbard 1990). Such critiques identify inherent gender biases in scientific narratives and processes, and see these as deeply rooted in a historical dichotomy dating back at least to Francis Bacon's figure of the pursuit of scientific knowledge as the domination of the female body of nature (Griffin 1978; Merchant 1980), a dichotomy that casts objectivity, reason and mind as male, and the body, emotion and nature as female. This underlying division does not only result in the exclusion of women from the practice of science, according to feminist critics, but shapes our very understanding of notions like nature, reason, science, labor, etc.

In the 1980s, such feminist critiques of science led to the conceptualization of alternatives to the objectification of nature in the form of romantic realist, organicist and eco-feminist approaches, which tended to celebrate "untouched nature", and to welcome, rather than resist, an engagement between women, nature and various forms of spirituality (Gaard 1993; Merchant 1996; Mies 1991; Shiva 1989). But while such approaches do incorporate the critique of nature as a resource for exploitation, they nonetheless produce two significant shortcomings according to

Haraway: totalizing theories which erase radical difference by claiming to speak for *all* women in the form of the first person plural; and an insistence on the organic – always defined in opposition to the technological – as the rallying point from which to resist patriarchic forms of domination. Haraway writes:

> There is nothing about being "female" that naturally binds women. There is not even such a state as "being" female, itself a highly complex category constructed in contested sexual scientific discourses and other social practices. Gender, race, or class consciousness is an achievement forced on us by the terrible historical experience of the contradictory social realities of patriarchy, colonialism, and capitalism. (1991: 155)

Haraway seeks here to oppose any attempt at grounding an essentialized unity or identity for women and to embolden women – formerly enclosed in discourses of the non-rational and the non-technical – to gain access to those spheres. Haraway's tactic is to stress the fictional nature of the ontological gender difference grounded in the affinity between women and nature: if nature and culture are constructs, as the very existence of the cyborg demonstrates, than there can be no metaphysical affinity between technoscience and gender difference. At the same time, she is using this feminist claim to undermine *all* essentialist categories, and above all that of nature:

> The theoretical and practical struggle against unity-through-domination or unity-through-incorporation ironically not only undermines the justifications for patriarchy, colonialism, humanism, positivism, essentialism, scientism, and other lamented -isms, but *all* claims for an organic or natural standpoint. (1991a: 157)

Thus, on an immediate level, the main contention of the Manifesto is its opposition to women's historic exclusion from science and technology. But the logic behind this opposition – that the binary categorization of nature/culture is arbitrary – is applicable beyond exclusively feminist concerns, to every field that is enframed by the nature/culture dichotomy, and it becomes clear that the underlying goal of the Manifesto is an attempt to conceptualize a *positive* engagement with technology. Herein lies the greatest novelty in Haraway's cyborg tale, as she calls on feminists and leftist intellectuals to embrace technoscience, rather than reject it in favor of a mythic, organic wholeness, or an anti-technology stance. Haraway writes,

> From *One-Dimensional Man* (Marcuse 1964) to *The Death of Nature* (Merchant 1980), the analytic resources developed by progressives have insisted on the necessary domination of technics and recalled us to an imagined organic body to integrate our resistance. (1991a: 154)

For Haraway, feminism and New Left socialism, the pillars of her intellectual upbringing, are too often immersed in a technophobic impasse that prevents them from seeing contemporary technoscience as a means of political action that might be used to alter the basis of life in positive ways. To reject technology and seek nostalgic recourse in an idea of nature or a repressed authentic humanity is seen as escapist.

In the radical posthumanist approach, such essentialist critiques that refer back to foundational narratives of the organic human or an uncontaminated nature are viewed as detrimental and untenable illusions. Rosi Braidotti (1996, 2002), for example, views techno-skepticism as a form of "nostalgic longing" for a supposedly better past that cannot adequately respond to the novel conditions and challenges of our present time. Mark Dery has argued that, "Neither nature nor the body exist anymore, in the Enlightenment sense; both are irredeemably polluted, philosophically speaking, in an age of human babies with baboon hearts and genetically altered mice with human genes" (1996: 245). And in a discussion about the patenting of transgenic organisms Haraway claims that "discourses of natural harmony, the nonalien, and purity [are] unsalvageable for understanding our genealogy ... It will not help – emotionally, morally or politically – to appeal to the natural and the pure" (1997: 62). For these theorists, there is hope in the void left by the collapse of overarching, foundational narratives, a void in which other partial and fragmented narratives concerned with identity and difference will be able to claim legitimacy.

The radical posthumanist call for a positive engagement with technoscience supposes the same constitutive nature of technology suggested by the substantive model by which technologies cannot be conceptualized as if they existed outside social contexts and as if they had, in turn, no role in shaping our engagement with the world. But in this approach this relationship need not be deterministic, rather, it should be understood as open, or reflexive (Graham 2002). The notion of reflexivity here embodies both the idea that technologies are the product of human creativity and that it is via our technologies that human ontology is realized. This is to say that technologies may be *determinative* of human experience, but they need not necessarily be *deterministic*. In this view technologies embody social biases or "politics" that are built in to them, but they also shape our social and political environment, often in very obvious ways, as television transforms consumption, the automobile reshapes the city and the clock synchronizes work. The reflexive model lies outside of the technophilic/techno-skeptic polarization, and implies that technology is both substantive and non-essentialist, constitutive and non-deterministic. A reflexive understanding of technology allows for its emergence from within social, political, and economic contexts and maintains that our technologies shape our engagement with the world. The reflexive view shares traits with both instrumentalism and substantivism: it agrees with instrumentalism that technology is in some sense controllable, and it agrees with substantivism that technology is value-laden.

4.2.2 Methodological Posthumanism: The Empirical Turn

Methodological posthumanism also generally rejects the dystopian inclinations of the earlier philosophers of technology. For methodological posthumanists, especially those of the newer generation of the philosophy of technology like Don Ihde (1990, 1993) and Peter-Paul Verbeek (2005; 2011), the sweeping claims of the classical philosophers of technology, what Carl Mitcham (1994) has called the "humanities" philosophy of technology, where technology is construed as a monolithic and deterministic phenomenon, prevents us from recognizing the myriad ways in which human contexts and values shape and constrain the use of technology. Empirical research into the development and use of specific technologies, they suggest, will lead to a move away from the transcendentalist approach and its emphasis on technology as alienation, towards a more nuanced view of technology/ies.

This "empirical turn" (Achterhuis 2001) in the study of technologies took its cue from developments that were taking place in the 1970s and 1980s in science studies, where a new emphasis on fieldwork, ethnographic interviews and archival research was favored over text-oriented and theoretical studies (Collins and Pinch 1982; Latour 1979; Lynch 1985; Knorr-Cetina 1981; Traweek 1988). These works opened up novel directions and methodologies for research that explored science as a practice and took particular phenomena, from technological milieus to laboratory culture to the role of the university and science policy, as its main research interests. In the mid-1980s, this novel inclination towards empirical based research was extended to the study of technologies and lead to richer conceptualizations of technology, as a political phenomenon (Winner 1980; Feenberg 1991), as a social activity (MacKenzie and Wajcman 1985; Bijker et al. 1987; Callon 1992) and as a cultural phenomenon (Borgmann 1984; Ihde 1990). These new approaches in science and technology studies rejected both the view that technology is a worldview or a historical necessity, and the view that technology is a neutral tool. Technological development is neither deterministic nor autonomous, nor does it follow a linear path from theory to application to attainment of ends. It is rather a highly contingent process, involving heterogeneous factors and influenced by social choices at every point of the way. Technologies here always bear the imprint of the social processes and social biases that have brought them forth and are built in to them, and are largely determined by the interpretive frameworks of the relevant social groups involved in their development. This means, in other words, that technology cannot have objective, intrinsic properties. Facts about a technology arise from the interpretations of relevant social actors, not from the technology itself; and technology cannot be analyzed by reducing it to its "conditions of possibility", but only in terms of concrete, technological artifacts.

An important consequence of this emphasis on specific technologies, on the materiality of technology, is the idea that technologies do not necessarily alienate humans from reality, but help shape their relationship with it. This is to say that technology may reduce certain forms of engagement with reality, but that it also creates new ones. This implies an understanding that the technological texture of the contemporary world is radically different than it was in Heidegger and Ellul's time, and that new ways of assessing technologies are necessary. Verbeek writes:

> Technology cannot be reduced without remainder to what underlies it. When Heidegger, for instance, conceives of technology as a dominating and controlling way of thinking and engaging with the world, and ultimately as a specific manner of world-disclosure or "being", he opens up an important perspective on technology. But this perspective is not sufficient to adequately analyze concrete technologies. To say that technologies *spring from* a certain manner of thinking and comporting oneself ... does not mean that such a manner of thinking and comporting is the only *allowable consequence* of using technologies. (2005: 8, emphases in original)

Verbeek argues that technology involves a lot more than the manipulation of objects and a reduction of reality to "standing reserve". A person who sends an email, he suggests, a passenger on a train, do not treat the addressee or the landscape as raw material. Thinking about technology requires questioning the ways in which the addressee and the landscape are present to the email sender and the passenger, and this requires what Verbeek calls the development of a "forward thinking", a means of assessing our engagements with different technologies and their presence in our lives.

This necessitates a move beyond the classical perspective in philosophy of technology that sees technology as something that estranges humans from reality, that diminishes the engagement of human beings with their environment. While loss of engagement might be a common aspect of modern technologies in light of their "disburdening" character, this is only one aspect of the implications of technology for the involvement of humans with their environment, and certainly not an inherent property of technology. Rather, Verbeek argues, amplification of engagement, or the creation of new forms of engagement is just as much a common aspect of modern technology, and the reduction of one form of involvement in the world as a result of the introduction of a new technology is most often accompanied by the creation of another form. This is illustrated in the critique Verbeek (2002) undertakes of Albert Borgmann's Heideggerian analysis of modern technologies in *Technology and the Character of Contemporary Life* (1984) and *Holding onto Reality* (2000). Borgmann's basic premise is that by relieving our efforts to accomplish things, technologies change the nature of our involvement in the world, and encourage a consumptive attitude. For example, the effort put into heating one's house in the past, including chopping wood, dealing with the hearth and then sitting around it with others, provided an engaged way of interacting with the world that is lost in the technology of central heating systems which produce a mere consumption of heat as a commodity.

But for every example of a disengaging technology that Borgmann provides, Verbeek cites a new form of engagement that emerges. So for example, against Borgmann's claim that the CD player does not provide us with the same type of access to reality that being present in a concert hall does, but replaces reality, Verbeek replies that the CD player has allowed music to become one of the most broadly enjoyed forms of art. Or, against Borgmann's claim that information technologies substitute reality with hyperreality and deliver reality as a commodity, Verbeek replies that information technologies mediate our environment and can enhance contact between people. Verbeek's development of a means of positively assessing contemporary technologies, without recourse to an uncritical technophilia, will be taken up in the mediated posthumanist perspective, and I shall return to it in later chapters. Suffice it to say here that both radical and methodological posthumanism implement a parallel shift that opens up the possibility of positive attitudes to contemporary technologies and biotechnologies that their theoretical origins did not allow for, though this shift indicates different directions for the radical and methodological approaches.

4.2.3   Technological Mediation: Heidegger, Ihde and Latour

The key concept in methodological posthumanism is technological mediation, which, like the notion of reflexivity in radical posthumanism, offers an alternative to instrumental and substantive models of technology, and of conceptualizing technology as either neutral or deterministic. Technological mediation implies that technologies play an active mediating role in the relationship between humans and their world. The notion of mediation also replaces alienation as the main concept for analyzing technology, insofar as technologies are no longer seen as artifacts that alienate humans from themselves and from nature, from an authentic way of being, but as offering one possible form of engagement with the world.

Heidegger's phenomenological analysis of the role of tools is often taken as a starting point for understanding the concept of technological mediation. In *Being and Time* (1962) Heidegger argues that tools should be understood as connections between humans and reality. Heidegger first claims that tools are not simply objects that have certain qualities, but that they are dependent on or relative to the context in which they are put to use. Tools are never understood as objects-in-themselves, but always in a complex field that is full of other involvements or cross-relations. Their intentionality is defined by the project being undertaken. In Heidegger's famous analysis of the hammer, for example, the hammer is understood in reference to the nails, the shingles, the carpenter and the task of nailing the shingles onto a roof. In terms of the user, the user's relationship to reality thus takes place through the use of the tool, which "withdraws" from the user's attention, or becomes "quasi-transparent" (Ihde 1990): the carpenter's attention is not directed at the hammer, but at the nail. The tool only really calls attention to itself when it "breaks down", when it no longer facilitates the relationship between user and world. Thus the tool is what Heidegger calls "ready-to-hand" (becoming "present-to-hand" when it breaks down) and it is through this "readiness-to-hand" that one's involvement with reality takes place, since this withdrawal reveals the environing world:

> Any work with which one concerns oneself is ready-to-hand ... also in the public world ... with the public world, the environing Nature is discovered and is accessible to everyone. In roads, streets, bridges ... our concern discovers Nature as having some definite direction. A covered railway platform takes account of bad weather. ... public lighting takes account of the darkness ... in a clock, account is taken of some definite constellation of the world system. (1962: 166, 181)

This is to say that technologies mediate our way of experiencing the world. They are not neutral intermediaries, but active mediators of how humans experience reality.

The notion of technological mediation has more recently been developed by Don Ihde and Bruno Latour, two of the leading figures in methodological post-humanist discourse. Heidegger's account of tool use is a phenomenological one, insofar as it is grounded in the understanding that humans and their world, or reality, are always interrelated, that humans are always directed at the world around them and that it is within this relationship that both humans and their world are (co-) constituted. Ihde also offers a phenomenological, or more precisely, what he calls a "postphenomenological" (1993b) approach to the study of technology.[2] Ihde discerns several relationships that human beings can have with technological artifacts: embodiment relations, hermeneutic relations, alterity relations and background relations.[3]

- In the *embodiment relation*, technologies, like Heidegger's ready-at-hand tools, are not objects of experience, but a means of experience. Technologies are incorporated by one's very bodily experience, and become extensions of the body. Ihde formalizes this relationship as: *(human-technology)* → *world*. An example of such a relationship is eyeglasses, which mediate one's vision, helping one perceive one's environment.
- In the *hermeneutic relation*, technologies provide access to reality not by being incorporated but by offering a representation of reality which then must be interpreted. These technologies engage with more linguistic and meaning-oriented capacities and draw attention to themselves all the while not being entirely present. This relation is formalized as: *human* → *(technology-world)*. Examples include instruments that offer readings or display gauges, which then require interpretation, such as the thermometer, which provides a value that establishes a relationship between humans and reality in terms of temperature.
- In *alterity relations* humans engage with technologies themselves as quasi-objects ("quasi" because they can never be present as a genuine other person). These technologies, such as toys, robots, or automatic machines, possess a certain degree of autonomy, and humans relate to the technology itself more than to the world through the technology. This relation is formalized as: *human* → *technology (−world)*.
- Finally, in *background relations* humans neither relate explicitly to the technology nor through the technology to the world, but the context of one's experience is shaped by the technology. This includes numerous technologies which are taken for granted and make up our environment in ways we are not conscious of (unless they "break down", in which case we are forcefully reminded of their existence), such as central heating and lighting. These technologies are part of the environment, which they help shape. This relation is formalized as: *I (−technology/world)*. According to Ihde, these relations form a continuum along which technologies are more or less conspicuous, but which in each case transform our experience of the world and transform ourselves in the process.

The work of Bruno Latour (1992, 1994) also offers important perspectives on the significance of mediation for analyzing technology. In *We Have Never Been Modern* (1993) Latour seeks to develop an "amodern" ontology that can overcome the deeply engrained separation or purification process of humans and

---

[2] Postphenomenology will be discussed in greater detail in Chap. 5.

[3] These relations appear as the basis of the phenomenology of technics Ihde has been developing over the last three decades. They are first clearly developed in *Technology and the Lifeworld* (1990).

non-humans undertaken by modernity, that is no longer sustainable, he claims in light of the growing number of interminglings between humans and non-humans. The asymmetrical treatment of humans and non-humans that is assumed in this separation, according to Latour, prevents us from developing a more realistic account of reality that presents both humans and non-humans as bound up with each other in a network of relations. "Actor-network theory" (ANT), an alternative framework developed by Latour and others (Callon and Law 1997; Latour 1992), assumes that the split between nature, society and artifacts is artificial, and that entities (human, non-human, textual or symbolic) do not have fixed boundaries but are defined by their relationships. Network and entity constitute each other here, allowing agency to be extended to all artifacts, since their existence always causes changes in behavior, routines and abilities.

In "Where are the Missing Masses?" (1992), Latour cites several examples that illustrate how artifacts can be deliberately designed to constrain and shape human action, decisions and mobility, from the seat-belt warning that alone can enforce the law on seat belts, to automatic doors that allow people to walk through them only at a certain speed, to bulky hotel key rings whose bulkiness prevents guests from putting them in their pocket and directs them to leave keys at the desk. As we shall see in Chap. 6, this also implies that artifacts can have normative or moral dimensions. ANT offers an understanding of how daily life is in many ways shaped by technologies. It offers a much more "symmetrical" perspective of the relations between humans and technological artifacts, insofar as there is no a priori distinction between human and non-human actors, rather both types of "actants" emerge from within the networks that exist between them. This is an anti-essentialist approach, since the existence of humans and non-humans does not emerge from within some essence, but from the relationships that are created in various, dynamic networks, in which the roles humans and non-humans play are equivalent. Neither actant has an essence and both are transformed in relation to one another.[4]

In Latour's symmetrical framework, the notion of mediation is paramount. It paves an alternative path between instrumentalism, or the belief that humans have any mastery over technology, and substantivism, or the belief that technology has some form of mastery over us. It can account for the idea that material artifacts are carriers of meaning that exert an influence on human action, and it can account for the hybrid actants that arise from the folding of humans and non-humans into each other in networks. To illustrate the notion of mediation, Latour analyzes the slogan developed by the National Rifle Association (NRA) that "Guns don't kill people; *people* kill people" (Latour 1999a: 176–180). In the event of a shooting, asks Latour, what role does the gun play? According to the opponents of gun sales, explains Latour, the gun changes everything, since it transforms an innocent citizen into a criminal. This is a substantivist account of gun use. According to the NRA, on the

---

[4] More recently Latour has turned away from the ANT framework, see Latour (1999b). Namely because each of the elements of ANT – actor-, network, theory and the hyphen between – have been too often misused, he explains.

other hand, the gun adds nothing to the action, it is a mere tool, a neutral vehicle that allows a human will to be carried out, more efficiently to be sure, but a will that preceded it; it does not modify its user in any meaningful way. This is an instrumental account. But neither of these approaches can account for the intermingling of humans and non-humans in the gun-citizen network that gives rise to the shooting event. The gun here, argues Latour, plays a mediating role that actively contributes to the way in which the event takes place.

Latour specifies four different aspects of technological mediation: translation, composition, reversible black-boxing and delegation. "Translation" indicates that the human/non-human or subject/object dichotomy is abandoned in the symmetrical approach, where a new, hybrid entity comprising both human intention and non-human function arises. In the example of the shooting, a human's "program of action", as Latour calls it, the intention to injure or kill, is mediated by an artifact's program of action, the gun's function of shooting, and the mediation gives rise to a new, translated program of action, the shooting event. Both actants are transformed in this network and responsibility for the shooting must be shared.

Next, "composition", which proceeds from the notion of translation, signifies that mediation always involves a number of actants working together: "Action is simply not a property of humans *but of an association of actants*" (Latour 1999a: 182, emphasis in original). In this sense, Latour explains, headlines such as "Woman flies to space" are misguided, because flying is not a property of that woman alone, but of an entire association of actants including humans, spaceships, launch pads, etc.

The third meaning of technological mediation is what Latour calls the folding of time and space, or "reversible blackboxing". The blending of humans and non-humans that occurs as a result of mediation is most often something that users are unaware of. Technological artifacts are always the product of more or less extensive networks that include manufacturers, material, distribution circuits, labor relations, etc. Usually, these networks are hidden to users, who are aware only of the finished product and its function as an independent object. The network of relations that produced the artifact is invisible and taken for granted. They relate to it, in Latour's words, like a "black box", the network only becomes visible when the artifact no longer functions, when it breaks down.[5] Latour uses the example of an overhead projector, the existence of which no one is really aware of during a projection until it ceases to function, when repairman rush to the scene and take it apart, and the individual function of each of its several parts comes to the fore, each a black box in itself.[6]

---

[5] This notion of the black box was first developed in science studies, where scientific theories were seen as black boxes that provide a truth about reality, but that conceal all the relations between scientists and the phenomena being investigated, all the problem-defining, experimentation and observation carried out to reach the theory.

[6] This is, of course, precisely the shift that Heidegger describes tools undergo from being "ready-to-hand" to "present-to-hand". Verbeek makes this point, and adds it to a list of examples in which Latour reveals his ignorance of Heidegger (Verbeek 2005: 158, note 7), at least the Heidegger of the tool analysis in *Being and Time* (1962). Worse, Verbeek explains, is the fact that in the chapter

The final and most significant aspect of mediation is "delegation". Techniques, argues Latour, have meaning, but they produce meaning through a type of articulation that overruns the distinction between signs and things, that he calls delegation. This implies that artifacts can exert an influence on their users not only as signs or carriers of meaning but as material things. Latour uses the example of a speed bump on a university campus to illustrate this. A driver's behavior is modified through the mediation of a speed bump insofar as she must slow down in order to protect her car's suspension. This involves a transformation in the driver's program of action, but also a change of the medium of expression. Drivers do not drive slowly because of a traffic sign or the presence of a policeman, but because of a concrete bump. A desired program of action, to make drivers slow down, is articulated through, or *delegated to*, an artifact (in this sense, as it is called in French, the speed bump is a "sleeping policeman"). Latour is careful not to use terms like "materialized", "engraved" or "reified" to account for this articulation because that would assume the imposition of a human will on innate matter, drawing us back into a subject/object dichotomy. Rather, delegation expresses the idea that a task has been passed on to an artifact which actively transforms programs of action, it is a "script" (Akrich 1992), in the sense that artifacts can prescribe their users how to act when using them. Both meaning and action have been displaced and translated in the speed bump.

Latour and Ihde develop different aspects of technological mediation. Namely, while Ihde focuses on a hermeneutic form of mediation that analyzes how technologies structure human perceptions and interpretations of reality, Latour focuses on a pragmatic form, that mainly analyzes action, or how specific technologies encourage people to act in certain ways. While the latter focuses on how human behavior is technologically mediated, the former is more interested in how artifacts can help shape human experience.[7] Still, these positions should be seen as complementing

---

on mediation in *Pandora's Hope*, Latour starts off with an attack on Heidegger's conception of technology and presents his analysis in opposition to Heidegger: "For Heidegger", Latour writes, "a technology is never an instrument, a mere tool. Does that mean that technologies mediate action? No, because we have ourselves become instruments for no other end than instrumentality itself" (1999a: 176). As Verbeek claims, Latour here relates to the Heidegger of "The Question Concerning Technology", and does not account for Heidegger's tool analysis as depicted above, which can be seen as a basis for the notion of technological mediation. Søren Riis, in his fittingly titled article "The Symmetry between Bruno Latour and Martin Heidegger" (2008), makes this critique of Latour even more clearly. "When looked at carefully", Riis argues, "Latour's examination of technical mediation stands out as a detailed reflection of Heidegger's studies" (285). There are other areas too where Latour's analyses are very Heideggerian, and resemble Heidegger's considerations on being and *Dasein*, such as his claim that "I want to situate myself at the stage *before* we can clearly delineate subjects and objects ... Full-fledged human subjects and respectable objects out there in the world cannot be my starting point; they may be my point of arrival" (Latour 1999a: 182). For this reason, in Chap. 6 that discusses posthuman approaches to subjectivity, I interpret Latour, as other methodological posthumanists insofar as subjectivity is concerned, as a continuation of Heidegger.

[7] For a discussion of the differences and similarities between ANT and postphenomenology see Verbeek (2005). Verbeek argues that the notion that these two views are incompatible arises mainly

each other, and of introducing a key novelty in technology studies: the notion that technology and technological artifacts must be understood in terms of mediation, between humans and their environment, between humans and other humans, and between humans and technology. Like the notion of reflexivity for radical posthumanists, this allows us to pave an alternative path beyond the view of technologies as either neutral tools that we can master or technologies as transformative devices that have some mastery over us. In other words, mediation is another way of overcoming the humanist separation between humans and technology that underlies the dominant posthumanist approaches.

4.3 The Prosthetic Nature of Human Being

The notions of reflexivity and technological mediation advanced by radical and methodological posthumanism do not only assume a different understanding of technologies than the essentialist understanding offered by instrumental and substantive models. They also have profound implications for conceptions of human nature and subjectivity. These are more explicit for radical than methodological posthumanism – as their differentiation along the historical-materialist and philosophical-ontological axis attested to in Chap. 2. And this presents a shortcoming for methodological posthumanism that I will return to in Chap. 6, as well as a more detailed discussion on the implications these approaches have for a posthuman subjectivity. Here, in this more general discussion on approaches to technology, I want to add a final term, prostheticity, that expresses these implications for human nature and subjectivity and that also frames and differentiates various approaches to technology.

4.3.1 Prostheticity and the Extension of the Self

Another means of understanding the difference between essentialist approaches to technology, be they instrumental or substantive, and non-essentialist approaches is the distinction between "supplemental" and "originary prostheticity". Supplemental prostheticity designates the type of relationship between humans and technology developed in classical and contemporary dominant understandings of technology as something that is ontologically distinct from nature, life, the human body and self, and is then "added on", acting as a supplement, as an appendage to these, *extending* their power.[8] In this view, technology can then be extolled or condemned when it

---

from Latour's anti-phenomenological stance and its alleged subjectivism, a critique that is precisely incorporated into Ihde's postphenomenological approach.

[8] I use the terms nature, life, human nature, body, self, subjectivity, etc., interchangeably here, since in any discussion of technology as extensive in a supplementary sense these terms interchangeably

serves or no longer serves but enslaves life and human nature. Originary prostheticity, on the other hand, refers to an understanding of the human body and self as already including prostheses as an integral part of its organization. In this view, technology is not extrinsic to human nature, rather it exists in relation to, and is dependent on, its technology. Technologies are not merely a grafting on or an appendage to the human body and self, but are more literally incorporated, assimilated or immersive. I borrow the term originary prostheticity from the French philosopher Bernard Stiegler (1998), who I will return to below, and who argues for an original bond between humanity and technology.

The distinction between supplemental and originary prostheticity is not self-evident, since in the technological register the term prostheticity already implies an "extension of self": not only in the literal sense of a prosthetic limb, but also in the more figurative sense that modern transportation becomes our prosthetic foot, glasses and telescopes our prosthetic eyes, and computers our prosthetic brain. The subtle but significant distinction that differentiates supplemental from originary prostheticity lies in two criteria. First, the extent of *boundedness* of that entity that comes into relation with technology – the self, nature or the body – prior to its encounter with technology, or prior to its "extension" (though the notion of priority is precisely what is undermined in originary prostheticity). And second, the extent of the *transformative* power on that entity that is attributed to technologies. Thus, in instrumental as well as substantive – i.e. essentialist – accounts, the encounter between humans and technologies is one of supplemental prostheticity because, as something that is "used" by, or on the contrary something that "impinges upon" humans, technology is ontologically separate from nature and humans, and improves or deteriorates some original condition. These approaches assume a norm of organic integrity where the human or nature is a point of origin. This is perhaps more obvious in substantive techno-skeptic approaches and traditional instrumental approaches than in the instrumental technophilia of liberal posthumanism, where technologies *do* seem to be integrated into human bodies, as examples of cognitive enhancement, virtual reality or wearable computers and sensory modalities illustrate, or vice versa, where humans seem to be integrated into a technology, as in the case of mind uploading. But liberal posthumanism retains an account of supplemental prostheticity of encounters with technology all the same: some initial, unified self remains intact and essentially unpenetrated by new technologies. This is the result, as suggested in Chap. 2, of the adoption of both a cybernetic model of the human *qua* biological organism and a humanist model of the human subject. Thus, if liberal posthumanism does not suppose a simple prosthetic framework that assumes a unified human body that is extended, it nonetheless assumes a unified and essential self or mind that subsists regardless of a tampering with or even doing away with the body.

Marshall McLuhan is usually credited with popularizing the notion that "tools are extensions of man". In his popular book *Understanding the Media: The Extensions of Man* (1974), McLuhan argued that following the mechanical age in

---

become the "other" of technology, while with regards to technology as extensive in an originary sense, the very opposition between technology and any of these terms is undermined.

which bodies had been extended in space, humans were approaching a final phase of extension symbolized by electronic communication technology. Media, he argued, are extensions of the human body and communication technologies are extensions of the human mind, "outerings" of what the body once enclosed. For McLuhan, technology is reflexive, insofar as it has the power to structure and restructure how human beings pursue their activities. The self is not just extended beyond its original location in the encounter with technology, it is transformed by it; namely, the actual structures of our mind are transformed by the different media we use: "the extension of any one sense", he writes, "alters the way we think and act – the way we perceive the world. When these ratios change, men change" (1967: 41). In this sense, McLuhan's work has helped problematize inner/outer or self/world distinctions in the direction of originary prostheticity. Indeed, his name often surfaces wherever cyborgs are concerned, and he has been a constant reference in the works of many radical posthumanists.[9]

But the notion that tools and technology have a prosthetic nature is nothing new. The idea that tools are extensions of the soul and of the body can already be found with Aristotle, who suggested that tools are inanimate slaves and slaves inanimate tools. In the *Eudemian Ethics* he writes that "the body is the soul's natural tool, while the slave is as it were a part and detachable tool of the master, the tool being a sort of inanimate slave" (1994: 1968). And in the *Politics*: "instruments are of various sorts; some are living, other lifeless; in the rudder, the pilot of the ship [the *kybernetes*] has a lifeless, in the look-out man, a living instrument; for in the arts [*technè*] the servant is a kind of instrument" (1996: book I, 1253b). For Aristotle, the extensive nature of tools is limited to the extension of the functions of the laboring body. Marx also adopts the notion of tool-use as an extension of the laboring body. In *Grundrisse*, he views technologies as extensions of the human will's domination over nature:

> Nature builds no machines, no locomotives, railways, electric telegraphs, self-acting mules, etc. These are products of human industry; natural material transformed into organs of the human will over nature ... They are organs of the human brain, created by the human hand. (Marx 1993: 706)

Writing at the end of the nineteenth century, Ernst Kapp (1877) also spoke of technique as an extension and prolongation of human organs, more specifically, of "organ-projection". Kapp conceived of technique as a kind of reflection according to which man projects himself into the outer world. Technical artifacts are developed in the image of his own organism following which man understands them as extensions of the latter. He writes:

> Since the organ whose utility and power is to be increased is the controlling factor, the appropriate form of a tool can be derived only from that organ. A wealth of intellectual

---

[9] McLuhan is often upheld as *the* theorist of cyberculture, and his readings of media as forms of human embodiment have been particularly significant for theories interested in the communities and identities made possible through web-based interactions, such as chatrooms and multiuser games. Arthur Kroker (1992) places McLuhan's extension thesis at the center of his analysis of technology and postmodernity. And Baudrillard has extended McLuhan's idea that "the medium is the message", namely in "The Implosion of Meaning in the Media" (1984).

creations thus springs from hand, arm and teeth. The bent finger becomes a hook, the hollow of the hand a bowl; in the sword, spear, oar, shovel, rake, plough and spade, one observes the sundry positions of arm, hand and fingers (in Mitcham 1994: 23–24).

The prosthetic nature of tool use has also been a theme in classical phenomenology. As we have seen, Heidegger's tool analysis described how tools are taken into the ways in which humans project themselves into work practices as they "withdraw" and become "ready-to-hand". Maurice Mearleau-Ponty's (1962) work on embodiment, although not specifically focused on technologies, also sheds light on prostheticity. For Merleau-Ponty, active, intentional bodily movements can incorporate objects or technologies, including them into subjective experience. His well-known examples include the blind man's cane and the woman's feathered hat, which extend the user or wearer's bodily experience beyond the outline of their biological body.

> The blind man's stick has ceased to be an object for him and is no longer perceived for itself; its point has become an area of sensitivity, extending the scope and active radius of touch and providing a parallel sight. In the exploration of things, the length of the stick does not enter expressly as a middle term, as an entity-in-itself; rather, the blind man is aware of it through the position of objects through it. The position of things is immediately given through the extent of the reach which carries him to it, which comprises, besides the arm's reach, the stick's range of action. (1962: 144)

The extension of bodily intentionality in Merleau-Ponty's examples is reminiscent of the cybernetic model discussed in Chap. 2, and the feedback loop in which information runs from a human body, to an environment and back again to form one system. As mentioned, it was one of the examples Gregory Bateson used to attempt to extend the new theoretical model of cybernetics and its transformation of the concept of boundaries and autonomy into the social sciences. In *Steps to an Ecology of Mind* he writes:

> It is not communicationally meaningful to ask whether the blind man's stick or the scientist's microscope are "parts" of the men who use them. Both stick and microscope are important pathways of communication and, as such, are parts of the network in which we are interested; but no boundary line – e.g., halfway up the stick – can be relevant in a description of the topology of this network. (1972: 251)

In the cybernetic perspective, information flows *through* the man-cane network, and it no longer makes sense to question where elements of a single system begin or end. Just like all of the examples given here, tool-use implies a certain prostheticity, which implies a certain extension of the body and self. But the question remains what kind of prostheticity these models assume.

4.3.2   Originary Prostheticity

The model of extension offered in these modern and pre-modern examples mostly suggest a supplemental prostheticity rather than an originary one. In Kapp's analysis for example, man gains *self*-knowledge through his technological culture. In this sense, the notion of "organ-projection" is more an attempt to naturalize the production of technological artifacts than to "technologize" human nature, or in other words, to establish the extensive nature of technology, rather than to argue for some original extensive nature of man. The precondition for man's *externalization* in his technological creations remains an idea of man as a unified identity, at least of an integrated unity of body and mind. Even in Marx's model, which can be called reflexive insofar as technology, namely machinery, is a socially structuring force which transforms consciousness, there is a concern to distance the techno- logical world from the natural realm that assumes an ontological distinction between the two. If Marx sees simple tools as extensions of the biological organ- ism, he also posits an abstract evolutive line in which machines evolve increasingly independently of their maker.

Supplemental prostheticity implies a process of addition which leaves largely intact the two categories of human body or self and technology that preceded their conjunction. For radical and methodological posthumanism, however, the human body/self and the construction of external and internal identity rely on techno- logy; prostheticity here is originary. Originary prostheticity (or "originary technicity") is the central premise of Bernard Stiegler's (1998) philosophy of technology, the idea that the human has always been technological. Stiegler, a contemporary of the radical and methodological posthumanists discussed here, draws on the perspective of the French paleoanthropologist André Leroi-Gourhan (1964), who argues for the coincidence of tool use and the appearance of the human. In *Technics and Time*, Stiegler undertakes a critique of Heideggerian and Habermasian approaches to technics, arguing that they both fail to think the fundamental co-emergence and co-dependency of technics and the human. Stiegler views the human as a lacking and undetermined being whose original incompletion is such that it is always already supplemented by technological prosthesis. A prosthesis, Stiegler claims,

> does not supplement something, does not replace what would have been there before it and would have been lost; it is added. ... the prosthesis is not a mere extension of the human body; it is the constitution of this body *qua* "human". ... It is not a "means" for the human but its end. (1998: 152–153)

Stiegler argues that the human specifically evolves through means *other than* life, through a coupling with the "exterior" evolution of technological objects. Instead of remaining committed to the essential distinction and even opposition between the human and the technical, a position which "forgets" the originary pros- thetic nature of humans, we should understand that technologies are the enabling condition, not obstruction, of human experience.

In originary prostheticity, the mode of encounter is no longer the meeting of one object and another, but of linkage, exchange and connection. Rosi Braidotti (1996), for example, speaks of a "mutual imbrication" of the technological and the human. "Far from appearing antithetical to the human organism and set of values", she writes, "the technological factor must be seen as co-extensive with and inter-mingled with the human". Elaine Graham, writing about the fear that our current complicity with technologies seems to give rise to, argues that humans have always co-evolved with their surroundings, tools and technologies. "To be human", she writes,

is already to be in a web of relationships, where our humanity can only be articulated – iterated – in and through our environment, our tools, our artifacts, and the networks of human and non-human life around us. (2004: 27)

This echoes the claim put forward in actor-network theory, that the elements that make up a network (be these human, non-human, natural or technical) mutually constitute each other. In this approach, the great oppositions of organic and technological (nature/culture) have been broken down into multiple networks that incorporate social, cultural and material relationships. Just as Haraway suggests that our integration with machines today has become as "natural" as using tools to extend the abilities of our bodies had become to pre-industrial men and women, and begs the question "Why should our bodies end at the skin, or include at best other beings encapsulated by skin? … machines can be prosthetic devices, intimate components, friendly selves" (1991: 178).

Catherine Waldby (2000) interprets originary prostheticity as "technogenesis", the idea that human beings are from the outset biotechnical networks, and that biology can only be made intelligible or have any value in light of the terms provided by technology. Waldby returns to Heidegger's notion of technics for the development of her thesis, using Samuel Weber's (1996) reading of "The Question Concerning Technology". Weber shows that, on Heidegger's own account, the distinction drawn between pre-modern and modern technics is untenable because all technics imply a "bringing-forth" of nature, the innermost principle of which is, writes Weber, "its impulse to open itself to the exterior, to alterity" (1996: 67). In this sense, technology-as-technics should be seen as one form of nature, or even as "more natural than nature itself" (67). And to be human, Waldby proceeds, has always meant to be "technological", indeed, there has never been a purely "organic" past that was subsequently "contaminated" by technicity.

Waldby uses this interpretation of Heidegger's understanding of the natural world as an open system as the basis for her radical analysis of the "Visible Human Project" (VHP), a venture undertaken by the National Library of Medicine that has made accessible three-dimensional recordings of human bodies that were dissected, photographed and turned into visual data files.[10] For Waldby the VHP is an emblematic instance of technogenesis (along with other dramatic biotechnical developments of our age, namely the Human Genome Project), that reveals the negotiable character of natural and biological entities and the ways in which they constantly engage with one another. Two elements are simultaneously at work in the VHP. As a new way to map and know the human, the VHP claims to define the human as a knowable species, a biological entity (its borders, its depths…). At the same time, Waldby

---

[10]The first of these images was made available on the National Library of Medicine's website, www.nlm.nih.gov/research/visible in November 1994.

argues, the transliteration and recording of the human as information implies a threat to any idea of the human as a stable, organic integrity:

> if human bodies can be rendered as compendia of data, information archives which can be stored, retrieved, networked, copied, transferred and rewritten, they become permeable to other orders of information, and liable to all the forms of circulation, dispersal, accumulation and transmission which characterize informational economies. (2000: 7)

The permeability of the human, or what Waldby also calls the "interramification" of lived bodies and technology, works to undermine the fantasy of bodily integrity, of the human as a purely organic, original entity whose limits can be specified. As we have seen, the anxiety caused by invasive technologies is thus interpreted as a reaction to this instability and as nostalgia for originary wholeness.[11] The categories of the body and the human both rely on the technologies that "invent" them. The VHP, Waldby explains,

> makes visible the extent to which the human is produced through encounters with those things that it putatively excludes – code, the corpse, its own endosoma, the computer. The VHP … confronts the human with those realms of production which it excludes from its self-image, showing the debt it owes to all those inhuman capacities which form its borders. The VHP lends an iconography to the idea of the human as synthetic, not a self-origin but rather the product of inestimable and incremental techno-bio-social processes. (161–162)

Similarly, but inspired by cognitive science, Andy Clark (1997, 2004) has developed a notion of extended mind, where body boundaries are treated as fluidly intermingled with technology. For Clark, the uniqueness of the human brain lies not in the idea that it is distinct from the rest of the natural order, a seat for the mind, but precisely in its ability to enter into deep and complex relationships with nonbiological constructs. Our notions of "mind", "person" and "self" are a result of this constant two-way traffic between "biological wetware" and tools and technologies. And our nature, Clark argues, is fundamentally cyborg. Clark's claim is an ontological one: we do not *become* cyborgs by the incorporation of wires, implants or silicon chips. We are born cyborgs by the very fact of our humanness, that is a result of an originary prostheticity. This is not just an argument for the "tool hypothesis" in cognitive development, by which the explosion of human intellectual abilities such as self-awareness, language and intelligence was triggered by tool use – but for the more radical claim that tool use may be at the origin of the emergence of the sense of self.[12]

The difference between supplemental and originary prostheticity hinges mostly upon the philosophical understandings of body and self that each one implies, rather than in the technologies in question themselves. A comparison between "real-life" cyborgs can illustrate this. Kevin Warwick, a professor of Cybernetics at the

---

[11] As Waldby remarks, the VHP itself is also infused with such dreams of wholeness: the two virtual bodies it has imaged are referred to as "Adam" and "Eve".

[12] Recently, some cognitive scientists have adopted this originary prosthetic perspective. In a series of experiments on macaques at the RIKEN Brain Science Institute in Japan for example, scientists concluded that treating simple tools as temporary extensions of the body induces a modification of body image that incorporates the tool, and that in evolutionary terms this in turn led to the gradual emergence of a sense of self more complex than the basic body image our evolutionary ancestors started out with. See (Ishibashi et al. 2000).

University of Reading in England, has undertaken a set of experiments known as "Project Cyborg" since the late 1990s in a bid to become a living cyborg.[13] In a first stage Warwick had a Radio Frequency Identification Device implanted in his arm that communicated with embedded sensors in the environment, allowing him to be identified by a computer in his building and to open doors and turn on lights. A second stage involved the neuro-surgical implantation of a device into the median nerves of his arm in order to link his nervous system directly to a computer, creating an integrated circuit. This allowed Warwick, for example, to be located in New York and, via the Internet, control a robot arm in Reading, England and to obtain feedback from sensors in his fingertips. During this same experiment, he also successfully connected ultrasonic sensors on a baseball cap and experienced a form of extra sensory input, as well as an electronic linkage between his own and his wife's nervous system. But Warwick's philosophical ruminations about such experiments do not venture into the realm of human subjectivity; the state of his cyborg being is based on supplemental prostheticity. For Warwick, these cyborg experiments will most immediately have an impact on research into physical disability, epilepsy and Parkinson's disease, and in a later age, on human evolution:

> In the future, I believe, we will be able to send signals to and from human and machine brains. We will be able to directly harness the memory and mathematical capabilities of humans. We will be able to communicate across the Internet by means of thought signals alone. Human speech and language, as we know it, will become obsolete. Ultimately, humans will become a lower form of life, unable to compete with either intelligent machines or cyborgs. (Warwick 2000: 1)

Subjectivity is not at stake in Warwick's type of cyborg: it remains whole, even as the body incorporates implants and is hooked up to computers. If any transformation is to happen, it will take place at the evolutionary scale.

For the Australian performance artist Stelarc, on the other hand, the very possibility of such corporeal experiments attests to the originary prostheticity of human body and selfhood. Indeed, Stelarc has been described by many theorists as exemplifying the posthuman condition.[14] Stelarc has been extending his body through performances since the late 1960s, beginning with suspension events and later through the use of robotics and computer technology that have involved attaching a "Third Hand" to his body, being remotely controlled by electronic muscle stimulators connected to the Internet in the "Movatar" project, and, more recently, grafting an organic ear onto his arm and projecting a 3D image of his own head onto a screen which viewers can converse with through a keyboard. Stelarc's work is based on the central idea of prosthetic selfhood, for which the human body is not a barrier, but a

---

[13] See www.kevinwarwick.org.

[14] See Joanna Zylinska's *The Cyborg Experiments* (2002b), which brings together a number of essays by leading theorists on Stelarc's works, Mark Dery's lengthy discussion on the artist in *Escape Velocity* (1996), and most recently, the anthology *Stelarc: The Monograph* (Smith 2007), with contributions by Jane Goodall, Arthur and Marilouise Kroker, Brian Massumi and William Gibson. Stelarc's website, www.stelarc.va.com.au provides graphic illustrations of his events as well as a many of his texts.

site on which to carry out its extensions. A traditional location of the self within a particular body is thus rendered meaningless, while the idea of "self-as-agent, skin-bounded or will-controlled" (Zylinska 2002a) becomes futile. Stelarc's cybernetic experiments aim to show that the body is not a container for subjectivity, but a network of additions, replacements and crossings, and that the evolutionary development of the body has always been intimately connected to technology:

> Ever since we evolved as hominids and developed bipedal locomotion, two limbs became manipulators. We have become creatures that construct tools, artifacts and machines. We've always been augmented by our instruments, our technologies. Technology is what constructs our humanity; the trajectory of technology is what has propelled human developments. I've never seen a body as purely biological, so to consider technology as a kind of alien "other" … is rather simplistic. (Stelarc 2002: 114)

Stelarc's events, seen in this way, are a performance of the notion of originary prostheticity. By introducing technologies into his body and onto its surface he stresses the nature of subjectivity as an extended operational system, as agency is dispersed over his body through networks of human and non-human entities.

4.3.3 Machinic Assemblages

The notion of originary prostheticity in one form or another has played a significant role in the school of French philosophical materialism developed by theorists like Gaston Bachelard (1934), Raymond Ruyer (1946), Georges Canguilhem (1975), André Leroi-Gourhan (1943, 1945, 1964) and Gilbert Simondon (1980). Beginning with Leroi-Gourhan's thesis on the co-evolution of human biology and technology, this rich tradition provides an important dimension of philosophical anthropology to the conceptualization of technics and technology. Working from an archaeological and paleo-anthropological standpoint, in a series of works Leroi-Gourhan develops a genealogy of human evolution along the two anthropological dimensions that typify the human species: language, the representative and symbolic capacity, and the brain/hand interface, a manual dexterity that is augmented by a representative reflexivity upon actions performed. He argues for an ancient and primordial alliance between these two, between *logos* and *techné*. He further suggests that from the first phase of evolution, the upright positioning of the human primate and tool use coincided, and that the adoption of tools can be seen as an expression of the exteriorization of biological functions. The early human did not suddenly begin using tools thanks to some "flash of genius", but acquired them in the gradual emission, or externalization of its body parts and brain. This resembles Kapp's theory of organ projection, and Leroi-Gourhan writes: "The whole of our evolution has been oriented toward placing outside of ourselves what in the rest of the animal world is achieved *inside* by species adaptation" (1964/1993: 236). But Leroi-Gourhan's theory implies an originary prostheticity, insofar as this projection begins as soon as humans can be called humans, and technology, once accessed, becomes itself a criteria of biological evolution. Material or technical culture here are inseparable from human nature.

This is also true of Georges Canguilhem's work on technology. In his 1947 lecture "Machine and Organism", for example, Canguilhem aims to undermine the conception of technology or the machine as the human other, claiming that "tools and machines are kinds of organs, and organs are kinds of machines" (1975: 143). Tools and machines, he suggests, are part of the living organism, if not parts that it was born with, than parts that it has made, so that they are part of *life*. In Canguilhem's biological philosophy of technology, machines are an extension of vitality, of the living force, they are a "projection of life" (Hacking 1998: 207). Of note is also Gilbert Simondon's (1980) thesis that technology cannot be reduced to a utilitarian function, but that it is a network of relations between tools and humans, tools and other tools, humans and other humans, tools and environments and all the cross-relations of mutual dependency and feedback that are produced amongst these. Technology is not "used" by active subjects to dominate an inanimate environment, it is the mediating agent between humans and the natural world that makes the illusion of an essentialist subject/object distinction possible.

It is in continuation of this line of thought that Gilles Deleuze and Félix Guattari also offer a model of originary prostheticity in their concept of the machine. In their terminology, a machine is any arrangement of heterogeneous parts, discontinuous alignments or linkages. Machines are fundamentally made of connection and disruption – of continuous flows that are interrupted and reconnected:

> A machine may be defined as a system of interruptions or breaks ... Every machine, in the first place, is related to a continual material flow ... that it cuts into. It functions like a ham-slicing machine, removing portions from the associative flow: ... the mouth that cuts off not only the flow of milk but also the flow of air and sound. (1977: 36)

For Deleuze and Guattari, reality is machinic. Such a claim obviously includes all things natural and hence human, and the machinic nature of humans is visible in the proliferation of connections among human and technical powers – so that a gun connecting flesh and metal at a distance is a machine, an eye that encounters a cinematic screen forms a machine, a hand that encounters the earth and acts as a tool is a machine. Nevertheless, "machine" seems to be a strange terminological choice, particularly when this is meant literally, not metaphorically, as they emphasize is the case (1977: 251).

But use of the term machine to illustrate originary prostheticity is neither random nor merely provocative, it is intentional and highly specific, emerging from the nature of machines as both non-organic and non-mechanical.[15] Traditionally, the machine or the mechanical system as a technological apparatus has been defined

---

[15] Earlier uses of the concept of the machine that Deleuze and Guattari claim debt to are Lewis Mumford's argument that society must be regarded as a machine and his description of certain ancient forms of empires as "megamachines" (see Deleuze and Guattari 1977: 251, 141), and Samuel Butler's fictional work, where he challenges the way in which lines are drawn between machinic and animal life (see Deleuze and Guattari 1977: 284). In "The book of the machines" section in his work *Erewhon*, Butler writes:

> Where does consciousness begin, and where end? Who can draw the line? Who can draw any line? Is not everything interwoven with everything? Is not machinery linked with animal life in an infinite variety of ways? The shell of a hen's egg is made up of delicate white ware and is a machine as much as an egg-cup. (1985: 199)

in contrast to the organism. Where the living organism is seen as enjoying a self-organizing capacity, in which the parts are both cause and effect of the whole, as displaying a finality, a unity, the machine is seen as lacking the power to reproduce and self-organize, its "cause" lying outside of it, in its designer. Deleuze and Guattari's machine takes from the "mechanical-as-non-organism" its lack of unity: the machine or assemblage follows no central or hierarchical order or organization; it is a fragmented aggregate whose parts do not constitute a unified whole (1977: 42).[16] But Deleuze and Guattari's machines are opposed to mechanical systems too (1977: 283–89), since they are not standardized systems that conform to a plan that applies principles. Unlike mechanical systems, the parts of a machine can be connected or disconnected in infinite ways, as long as desire is made to flow. What's more, machines, contrary to mechanisms, work best by "breaking down", since they are constituted as a system of interruptions and breaks – they involve flows that are cut into and constantly redistributed without ever being successfully organized. In Deleuze and Guattari's form of realism, where the animate and the inanimate all have the same ontological status, the mechanical/organic opposition is redundant, and machine is pitted against either of these.

The machine, and the assemblage as machinic, does not describe a technological device or tool, but refers to a productive connection of elements. Thus, it is precisely in the seemingly inappropriateness of this highly technological term that the notion of originary prostheticity is best expressed. In the model of machinic realism it is not a question of the supplemental prosthetic encounter of a *meeting* between two objects, but of linkage, exchange and connection. Technology does not *meet* a body, rather the matter, flows and intensities of the biological connect with other matter and flows of the technological, and vice versa, the elements of an assemblage cannot be traced back to original unified components. In any given assemblage, furthermore, these exchanges can be distributed in different ways, and insofar as human bodies can forge an endless number of new connections, they remain open.

4.4 Conclusion

The notion of originary prostheticity is another means of expressing what a non-essentialist, and so a non-humanist, model of technology signifies. It implies that the human exists in relation to and is dependent on its technologies; that the human emerges as a result of this relationship. Originary prostheticity calls for a radical questioning of the limits of humanity. In this sense, originary prostheticity is also a

---

[16] In *A New Philosophy of Society* (1985) Manuel DeLanda carries out an excellent discussion of this aspect of assemblages and their contrast to Hegelian totalities. Organic wholes, he explains, are characterized by "relations of interiority" such that all component elements are identified by their functioning in constituting the whole, and that, detached from the whole, the elements no longer exist. Assemblages, by contrast, are characterized by "relations of exteriority", such that the identities of component elements are not reducible to the relations in which they find themselves at any time, i.e., they are assumed to self-subsist. The assemblage thus possesses synthetic or emergent properties.

basis for radical and methodological posthumanist models of technology. As we have seen in this chapter, these approaches develop an alternative framework that overcomes the pessimism and transcendentalism of the critical approaches represented by classical philosophy of technology as well as the essentialism inherent in instrumental and substantive models of technology that inform dystopic and liberal posthumanism. In this alternative the dualist paradigm of humanism according to which humans are autonomous, independent and unique entities, distinctly separate from their world and from their technologies, simply cannot account for the deep intimacy, the intricate enmeshing between humans and technologies that is an integral part of human experience.

In order to take account of this messiness, technology must be approached in terms of specific artifacts, that each have varying degrees of influence and varying degrees of transformative powers on human action and experience. The creation of new analytical tools is necessary – a task achieved by radical posthumanism with the notion of reflexive technology, and to an even greater extent by methodological posthumanism with the notion of technological mediation, be this practical or hermeneutic. In this framework, technology is a specifically human relationship, a mode of active relation to the world. The *logos/techné* alliance is seen as originary; it is at the origin of what it means to be human. Technology cannot be de-humanizing, a force that alienates humanity from itself in this sense, because it is at work in the very humanization process of the human. From here, both radical and methodological posthumanism argue for more positive conceptualizations of technology than previous critiques allowed for. For the former, as we shall see in greater detail in later chapters, this implies a celebration of the political potential inherent in new technologies to overcome some of the most detrimental effects of modernity. For the latter, this means conceptualizing the ambivalent status of technology, which may lead to a loss of involvement of humans in their environment in some instances, but also amplifies and creates new forms of engagement.

References

Achterhuis, H. (2001). *American philosophy of technology: The empirical turn*. Bloomington: Indiana University Press.

Akrich, M. (1992). The de-scription of technical objects. In W. Bijker & J. Law (Eds.), *Shaping technology/building society* (pp. 205–224). Cambridge, MA: MIT Press.

Aristotle. (1994). Eudemian Ethics. In J. Barnes (Ed.), *The complete works of Aristotle, vol. 2*. Princeton: Princeton University Press.

Aristotle. (1996). Politics. In S. Everson (Ed.), *Aristotle, the politics and the consitution of Athens*. Cambridge: Cambridge University Press.

Bachelard, G. (1934). *Le nouvel esprit scientifique*. Paris: Les Presses Universitaires de France.

Bateson, G. (1972). *Steps to an ecology of mind*. New York: Ballantine Books.

Baudrillard, J. (1984). *Simulacra and simulation* (trans: Glaser, S.F.). Ann Arbor: University of Michigan Press.

Bijker, W. E., Hughes, T. P., & Pinch, T. (Eds.). (1987). *The social construction of technological systems: New directions in the sociology and history of technology*. Cambridge, MA: MIT Press.

Borgmann, A. (1984). *Technology and the character of contemporary life*. Chicago: University of Chicago Press.

Borgmann, A. (2000). *Holding on to reality: The nature of information at the turn of the millenium*. Chicago: University of Chicago Press.

Braidotti, R. (1996). Cyberfeminism with a difference. http://www.let.uu.nl/womens_studies/rosi/cyberfem.htm. Accessed 13 June 2013.

Braidotti, R. (2002). *Metamorphoses: Towards a materialist theory of becoming*. Cambridge: Polity Press.

Butler, S. (1985). *Erewhon*. Harmondsworth, Middlesex: Penguin. Original edition, 1872.

Callon, M., & Latour, B. (1992). Don't throw the baby out with the Bath school! A reply to Collins and Yearley. In A. Pickering (Ed.), *Science as practice and culture* (pp. 343–368). Chicago: Chicago University Press.

Callon, M., & Law, J. (1997). After the individual in society: Lessons on collectivity from science, technology and society. *Canadian Journal of Sociology, 22*(2), 165–182.

Canguilhem, G. (1975). Machine et organisme. In *La Connaissance de la Vie*. Paris: Vrin.

Clark, A. (1997). *Being there. Putting brain, body and the world together again*. Cambridge, MA: MIT Press.

Clark, A. (2004). *Natural-born cyborgs: Minds, technologies, and the future of human intelligence*. Oxford: Oxford University Press.

Collins, H., & Pinch, T. (1982). *Frames of meaning*. London: Routledge.

Dery, M. (1996). *Escape velocity: Cyberculture at the end of the century*. New York: Grove.

Deleuze, G., & Guattari, F. (1977). *Anti-oedipus: capitalism and schizophrenia* (trans: Seem, M., Lane, H. R., & Hurley, R.). New York: Viking Press. Original edition, 1972.

Dessauer, F. (1927). *Philosophie der Technik: Das Problem der Realisierung*. Bonn: F. Cohen.

Dewey, J. (1929). *Quest for certainty: A study of the relation of knowledge and action*. London: George Allen & Unwin.

Ellul, J. (1965). *The technological society* (trans: Wilkinson, J.). New York: Vintage.

Feenberg, A. (1991). *Critical theory of technology*. New York: Oxford University Press.

Gaard, G. (Ed.). (1993). *Ecofeminism: Women, animals, nature*. Philadelphia: Temple University Press.

Ortega y Gasset, J. (1939). *Meditación de la Técnica*. Madrid: El Arquero.

Graham, E. L. (2002). *Representations of the post/human: monsters, aliens and others in popular culture*. New Brunswick: Rutgers University.

Graham, E. L. (2004). Post/human conditions. *Theology and Sexuality, 10*(2), 10–32.

Griffin, S. (1978). *Woman and nature: The roaring inside her*. New York: Harper & Row.

Habermas, J. (1968). *Knowledge and human interest*. Cambridge: Polity Press.

Hacking, I. (1998). Canguilhem amid the cyborgs. *Economy and Society, 27*(2&3), 202–216.

Haraway, D. (1991). A cyborg manifesto: Science, technology, and socialist-feminism in the late twentieth century. In D. Haraway (Ed.), *Simians, cyborgs and women: The reinvention of nature* (pp. 149–181). New York: Routledge.

Haraway, D. (1997). *Modest_Witness@Second_Millenium. FemaleMan©_Meets_Oncomouse™: Feminism and Technoscience*. New York: Routledge.

Harding, S. (1986). *The science question in feminism*. Ithaca: Cornell University Press.

Harding, S. (1991). *Whose science? Whose knowledge? Thinking from women's lives*. Ithaca: Cornell University Press.

Heidegger, M. (1962). *Being and time* (trans: Macquarrie, J., & Robinson, E.). New York: Harper & Row. Original edition, 1927.

Heidegger, M. (1977). The question concerning technology. In D. Farell Krell (Ed.), *Martin Heidegger: Basic writings* (pp. 287–317). New York: Harper & Row.

Hubbard, R. (1990). *The politics of women's biology*. New Brunswick: Rutgers University Press.

Ihde, D. (1990). *Technology and the lifeworld: From garden to earth*. Bloomington: Indiana University Press.

Ihde, D. (1993a). *Philosophy of technology: An introduction*. New York: Paragon.

Ihde, D. (1993b). *Postphenomenology: Essays in the postmodern context*. Evanston: Northwestern University Press.

Ishibashi, H., Sayaka, H., & Atsushi, I. (2000). Acquisition and development of monkey tool-use: Behavioral and kinematic analyses. *Canadian Journal of Physiology and Pharmacology, 78*(11), 958–966.

Jaspers, K. (1933). *Man in the Modern Age* (trans: Paul, E., & Paul, C.). London: Routledge. Original edition, 1931

Jonas, H. (1979). *The imperative of responsibility: In search of ethics for the technological age.* Chicago: University of Chicago Press.

Kapp, E. (1877). *Grundlinien einer Philosophie der Technik.* Braunschweig: Westermann.

Keller, E. F. (1985). *Reflections on gender and science.* New Haven: Yale University Press.

Keller, E. F. (1995). *Refiguring life: Metaphors of twentieth century biology.* Colombia: Colombia University Press.

Knorr-Cetina, K. (1981). *The manufacture of knowledge.* New York: Pergamon.

Kroker, A. (1992). *The possessed individual.* New York: Macmillan.

Latour, B. (1992). Where are the missing masses? Sociology of a few mundane artefacts. In W. E. Bijker & J. Law (Eds.), *Shaping technology/building society: Studies in sociotechnological change* (pp. 225–259). Cambridge, MA: MIT Press.

Latour, B. (1993). *We have never been modern.* Cambridge: Harvard University.

Latour, B. (1994). On technical mediation: philosophy, sociology, genealogy. *Common Knowledge, 3,* 29–64.

Latour, B. (1999a). *Pandora's hope: Essays on the reality of science studies.* Cambridge, MA: Harvard University Press.

Latour, B. (1999b). On recalling ANT. In J. Law & J. Hassard (Eds.), *Actor-network theory and after* (pp. 15–25). Oxford: Blackwell.

Latour, B., & Woolgar, S. (1979). *Laboratory life: The social construction of scientific facts.* London: Sage.

Leroi-Gourhan, A. (1943). *Evolution et techniques I: L'homme et la matière.* Paris: Albin Michel.

Leroi-Gourhan, A. (1945). *Evolution et techniques II: Milieu et technique.* Paris: Albin Michel.

Leroi-Gourhan, A. (1964). *Le geste et la parole.* Paris: Albin Michel. English Edition: Leroi-Gourhan, A. (1993) *Gesture and speech* (1993). (trans: Bostock Berger, A.). Cambridge: MIT Press.

Lynch, M. (1985). *Art and artifact in the laboratory.* London: Routledge.

MacKenzie, D., & Wajcman, J. (Eds.). (1985). *The social shaping of technology.* Milton Keynes: Open University Press.

Marcuse, H. (1964). *One-dimensional man: Studies in the ideology of advanced industrial society.* Boston: Beacon.

Marcuse, H. (1998). Some social implications of modern technology. In D. Kellner (Ed.), *Technology, war and fascism: Collected papers of Herbert Marcuse* (Vol. 1). London: Routledge.

Marx, K. (1993). *Grundrisse* (trans: Nicolaus, M.). London: Penguin Books. Original edition, 1857.

McLuhan, M. (1974). *Understanding media: The extensions of man.* London: Abacus.

McLuhan, M., & Fiore, Q. (1967). *The medium is the massage: An inventory of effects.* New York: Bantam.

Merchant, C. (1980). *The death of nature: Women, ecology, and the scientific revolution.* New York: Harper and Row.

Merchant, C. (1996). *Earthcare: Women and the environment.* London: Routledge.

Merleau-Ponty, M. (1962). *Phenomenology of perception* (trans: Smith, C.). London: Routledge & Paul Kegan. Original edition, 1945.

Mies, M. (1991). *Patriarchy and accumulation on a world scale: Women in the international division of labour.* London: Zed Books.

Mitcham, C. (1994). *Thinking through technology.* Chicago: Chicago University Press.

Mumford, L. (1934). *Technics and civilization.* New York: Harcourt, Brace and World.

Riis, S. (2008). The symmetry between Bruno Latour and Martin Heidegger: the technique of turning a police officer into a speed bump. *Social Studies of Science, 38*(2), 285–301.

Ruyer, R. (1946). *Eléments de psycho-biologie.* Paris: Presses Universitaires de France.

Shiva, V. (1989). *Staying alive: Women, ecology and development.* London: Zed Books.

Simondon, G. (1980). *On the mode of existence of technical objects* (trans: Mellamphy, N.). London: University of West Ontario.

Smith, M. (Ed.). (2007). *Stelarc: The monograph*. Cambridge, MA: MIT Press.

Stelarc (2002). Probings: An interview conducted by J. Zylinska and G. Hall. In J. Zylinska (Ed.), *The cyborg experiments: The extensions of the body in the media age* (pp. 114–130). London: Continuum.

Stiegler, B. (1998). *Technics and time, 1: The fault of epimetheus*. Stanford: Stanford University Press. Original edition, 1994.

Traweek, S. (1988). *Beamtimes and lifetimes*. Cambridge, MA: Harvard University Press.

Verbeek, P. P. (2002). Devices of engagement: On Borgmann's philosophy of information and technology. *Techné, 6*(1), 69–92.

Verbeek, P. P. (2005). *What things do: Philosophical reflections on technology, agency and design*. University Park: Penn State University Press.

Verbeek, P. P. (2011). *Moralizing technology: Understanding and designing the morality of things*. Chicago: Chicago University Press.

Waldby, C. (2000). *The visible human project: Informatic bodies and posthuman medicine*. London/New York: Routledge.

Warwick, K. (2000). I Want to be a Cyborg. *The Guardian*, (January 26).

Weber, S. (1996). *Mass mediauras: Form, technics, media*. Sydney: Power Publications.

Winner, L. (1980). Do artifacts have politics? *Daedalus, 109*, 121–136.

Zylinska, J. (2002a). The future… is monstrous: Prosthetics as ethics. In J. Zylinska (Ed.), *The cyborg experiments: The extensions of the body in the media age* (pp. 214–236). London/New York: Continuum.

Zylinska, J. (Ed.). (2002b). *The cyborg experiments: The extensions of the body in the media age*. London/New York: Continuum.

Chapter 5
From Molar to Molecular Bodies: Posthumanist Frameworks in Contemporary Biology

Abstract This chapter steps back from the technological realm and takes a look at how the humanist dualist paradigm is also being challenged in current biological research, particularly in molecular biomedicine and evolutionary biology. A possible shift in these disciplines attests to this, from a "molar" formulation of the body or organism, understood as a self-contained, unified organic whole, distinct from its environment, to a "molecular" body or organism, understood as a fragmented assemblage made up of transferable and translatable parts that depends much more on interactions with its surroundings. This biological form of originary prostheticity complements its anthropological counterpart that was articulated in Chap. 4.

Keywords Molecularization • Evolutionary biology • Molarity • Symbiosis • Network

The various approaches to technology that were analyzed in the previous chapter gave rise to the significant distinction between supplementary and originary prostheticity: a reiteration in terms of technology of the more general distinction between humanist and non-humanist approaches to emerging biotechnologies.[1] In this framework, supplementary prostheticity presupposes a body that engages with technology as a coherent unified whole, as a bounded organism which is "supplemented" by technology, for better or worse. This type of body, in line with Deleuze and Guattari's (1987) distinction between the *molar* and the *molecular*, can also be called a molar body: a body which in its encounter with technology presumes a norm of organic integrity that maintains an ontological divide between itself and its technological other. The dualist paradigm of humanism is maintained here on the level of corporeality. Originary prostheticity presupposes a different formulation of the body, whose organization already includes and depends on

---

[1] This chapter is inspired by the article "The Missing Link. How Biology can Help Philosophy of Technology Complete its Ontological Shift" (2013). *Tijdschrift voor Filosofie, 75*(1), 121–145.

T. Sharon, *Human Nature in an Age of Biotechnology: The Case for Mediated Posthumanism*, Philosophy of Engineering and Technology 14, DOI 10.1007/978-94-007-7554-1_5, © Springer Science+Business Media Dordrecht 2014

"exterior" technological objects. This "molecular" body is understood more as a fragmented, machinic assemblage, made up of transferable and translatable parts, an open nexus that can be composed, decomposed and recomposed via its interactions with its surroundings.

These configurations of the body, molar and molecular, will be the focus of this chapter. But here we will take a detour away from the realm of technology to the domain of biology, namely molecular biomedicine and evolutionary biology, focusing, so to speak, on the strictly "bio-" element in emerging biotechnologies. Indeed, some recent trends in biology can be seen as mirroring the same move beyond the subject/object or inner/outer opposition that informs methodological and radical posthumanism, whereby a molar configuration of the body or the organism is being replaced by a molecular one, that conceives the body or organism as an open-ended network whose boundaries are in constant flux and negotiation with its environment. Such research suggests that the notion that biological organisms have fixed spatial and temporal boundaries does not correspond to how the natural world works, and that organisms are made up of elements and mechanisms that can be isolated and mobilized, that they are a result of an interaction with their environment, and that they often emerge from processes of hybridization and inter-species exchange. In other words, the originary prostheticity that methodological and radical posthumanism acknowledge as an inherent feature of the encounter between humans and technology is also being identified here, on very basic, rudimentary levels of biological life.

Three examples from contemporary biology will be discussed. First the view generated in molecular bioscience of a "recombinatory" body made up of flexible and mobile elements of genetic information that can be transferred between bodies and between species. Second, new models of dynamical systems in evolutionary biology, which are promoting a view of the organism as an open system that actively participates and interacts with its environment. And third, molecular phylogeny, which argues for a view of species and organisms as the results of endosymbiotic fusions and genetic flux between domains of life facilitated by the cross-taxa dispersal of viruses. These developments may seem only remotely related to the discussion on posthumanism. But such non-humanist stirrings in contemporary biology can complement and support efforts to develop non-humanist models for understanding the implications of emerging biotechnologies for what it means to be human.

5.1 From the Clinical Gaze to the Molecular Gaze

5.1.1 Modern Biomedicine and the Molar Body

In *The Birth of the Clinic* (1973) Michel Foucault traces a number of transformations in medical thought and practice at the end of the eighteenth and the beginning of the nineteenth centuries that he claims led to the development of modern medicine. I will return to some of these transformations in greater detail in Chap. 8, namely to the shift from a medicine of "surfaces" and classifications to a medicine of "depth" and its implications for notions of subjecthood in the context of genomics and neuroscience. But most significant for a discussion on models of posthuman corporeality is Foucault's claim that these transformations introduced a new notion of the body that framed the models of biological and medical thought that developed from this period onwards. According to Foucault, eighteenth-century medicine was based on identifying diseases in terms of their resemblance and difference as located within a general table of diseases. Within this classificatory schema, diseases were treated as natural kinds, each caused by a specific agent, and situated within families, genera and species. In this classificatory space, disease was construed as a sum of the trail of symptoms that marked its passage through the inside and outside of the body, and the individual patient was relevant only insofar as disease moved throughout its body and relied on its organs for support. But towards the end of the eighteenth century, Foucault argues, the creation of new medical ways of seeing and knowing the body reconstrued illness as a specific anatomical lesion located in the analyzable three-dimensional structure of the body, thus shifting the focus of medical inquiry to the individual body and its organs as the space of illness. He writes,

> Disease is no longer a bundle of characters disseminated here and there over the surface of the body and linked together by statistically observable concomitances and successions. ... It is no longer a pathological species inserting itself into the body wherever possible; it is the body itself that has become ill. (1973: 136)

The relocalization of illness in the body marks the passage from a medicine of species and classifications to a medicine of depth, organs and functions, in which the body became the object of what Foucault calls the clinical gaze: the ability to see through the density of the corporal tissues to the hidden source of disease. This clinical gaze encompassed both a system of knowledge that equated illness to the underlying pathological lesion and a new method of clinical practice that allowed for the access to and visualization of the body, via post-mortem dissection, new scrutinizing devices such as the stethoscope and later on x-rays and microscopes, and the inscription of the body in the anatomical atlas. Under the clinical gaze, the body became a static entity that could be penetrated in order to find the "real" cause of disease. The modern clinical body hereafter became a bounded living organism, made up of functionally connected components (such as organs and tissues) and internal systems and processes (such as feedbacks, rhythms, and circulations), an organic and functional unity that is at constant risk of disruption by disease.

This "molar" body, the body as a systemic whole, that took shape in modern clinical medicine, can be seen as the dominant model of the body throughout the nineteenth and twentieth centuries in a number of disciplines. Thus in the clinical model, health relies on the stability or homeostasis of the body's various internal organic systems (the digestive, the reproductive, the endocrine, or the cardiovascular). The living body here is an organic unity, and is constantly under threat. This image is conveyed by the militant terminology that is so often employed in the discussion on disease: healthy bodies are "invaded" and "attacked", by diseases that are "enemy" and "foreign" to them. Recurrent metaphors here include the body as a battleground, the body under siege, medicine as a weapon or, when it leads to a cure, a magic bullet.[2] This molar body is also at the center of the zoocentric model of modern biology, which offers a picture of the body as a closed unity based on the notion of the organism as a self-regulating and self-generating entity that has been advanced since classical Darwinism and up to autopoeitic theory. Thus we think of evolution in terms of the functioning of fully determinate and discrete units (whether these are organisms, or more recently, genes). The Oedipalized body of psycho-analysis is also construed as a molar body. Psychoanalysis describes how bodily functions and drives are integrated into a coherent unified body through the imposition or acquisition of a particular social inscription, resulting in the formation of consciousness and the integration process of a fragmented body. Indeed, the failure to achieve this cohesiveness results in neurotic symptoms which are the manifestations of a ruptured unity.

5.1.2 Processes of Molecularization

But in contemporary biology and biomedicine, the model of the molar body that has helped shape so much theory about corporeality over the past two centuries is increasingly being displaced by a more molecular model, in which the body is no longer perceived as a self-contained, unified organic whole, but as an assemblage of discrete and transferable elements. Beginning in the 1930s, and more considerably in the 1950s, biology began to visualize life at the molecular level, in terms of submicroscopic developments (Kay 1993). As a number of scholars of science, technology and medicine have noted, the "molecularization" of the life sciences has had significant repercussions across multiple disciplines (Abir-Am 1985; Chadarevian and Kamminga 1998; Kay 1993; Rabinow 1992; Rose 2001). Nikolas Rose, for example, has argued at length that the repercussions of the process of molecularization are not merely a matter of framing explanations at a more detailed, molecular level, or even of using instruments that are made at the molecular level. Molecularization, rather, implies much deeper transformations concerning a "reorganization of the gaze of the life sciences", along with "their institutions, procedures, instruments, spaces of operation and forms of capitalization" (Rose 2001: 13). Like Foucault's clinical gaze, which it supplements if not comes to replace, the molecular gaze consists of both a molecular knowledge of life – that understands life in terms of genes, proteins and enzymes, and new methods and techniques for access to and intervention at the molecular level – from

---

[2] For a detailed discussion of such metaphors, namely in the context of cancer, see Susan Sontag (1977) and Richard Gwyn (2002). Sontag comments on the fact that the same vocabulary is used in reference to cancer, aerial warfare and science fiction. Cancer cells invade the body, patients are bombarded with toxic rays and chemotherapy is construed as chemical warfare.

techniques of gene cutting and splicing, to polymerase chain reaction and DNA diagnostic tools. And just as a new notion of the body emerged with the formation of the clinical gaze, a new molecular notion of the body has begun to emerge from within the molecularization of life. This molecular body differs in two main ways from the molar body of modern biomedicine: it is conceived on a different scale, and secondly, it is fragmented into transferable elements of genetic information.

The most notable difference between the molecular and molar configurations of the body is obviously a difference of scale. While life in the framework of the molar body was conceptualized at the level of organs, tissues and blood flows, on a 1:1 ratio, life in the molecular body is a phenomenon that takes place at the submicroscopic region. As Foucault (1973) argues, the clinical gaze aimed at bringing the hidden organic depth of the living individual to light, by transcribing invisible bodily facts into auditory representations (stethoscope) or into visual representations, either graphic (photography, radiography) or numeric (thermometry, sphygmography). With the use of increasingly sophisticated imaging systems today, such as ultrasound, computer tomography (CT) and magnetic resonance imaging (MRI), this interiority is provided via fuller visual representations, delivered in three dimensions, point by point, and in multiple slices, and the activity of neurons, the velocity of blood flow, the concentration of chemicals, can all be elicited from any viewpoint. The molecular gaze, then, can be seen as continuing this aspiration to penetrate the interiority of the body at ever deeper levels, only the localization of disease now takes place on a deeper, or more detailed level. This is not to say that disease diagnosis and treatment does not continue to work on the molar level as well, in terms of pathologies or organs or systems. But increasingly, diseases and disorders from cancer subtypes to different forms of depression are being perceived as having molecular bases, and therapeutic research is carried out at the molecular level where molecular agents and mechanisms can be manipulated.[3]

This difference of scale between molar and molecular bodies is significant insofar as the localization of disease shifts once more. But it does not necessarily indicate more than a difference in degree, since it can be understood as the same *type* of gaze on different anatomical levels, enabled by the greater magnifying power of newer technologies. However, this shift to the molecular also entails a difference in kind. The techniques that make possible the molecularization of life do not only assume that molecular entities and mechanisms can be identified and isolated in greater and greater detail – they also assume that they can be manipulated, mobilized and recombined. It is this aspect of the molecular model that indicates a real shift from the molar.

This understanding, of the recombinatory nature of molecularized life, owes much to the informational language that has framed molecular biology and genetics since the 1920s. As briefly outlined in Chap. 2, the informational model that early

---

[3] The new generation of "targeted therapies" in cancer research, for example, bind to signaling molecules on tumors in order to disrupt cell growth signals (Genentech's Herceptin for breast cancer, Pfizer's Sutent for kidney cancer), or to hit specific molecular components of the immune system (Medarex's Ipilimumab).

cybernetic theory developed attempted to provide a universal language of information into which the basic elements of all systems could be translated. In the context of biology, this new widespread emphasis on information as the main frame of reference led to a reconceptualization of heredity as information transfer, and of biological material, namely genes, as information storage and retrieval systems (Canguilhem 1994; Kay 2000; Keller 2000). The other radical novelty of this informational model, as mentioned, was to dissociate information from the material substrate that it is "located" in, so that a distinction could be established between information or meaning and the physical channel that transfers or conveys it.

In the biological setting, this meant that the biological organism was translated into problems of genetic coding and readout, insofar as biological components, such as tissues, cells, DNA fragments and genes, were conceived of as sequences of digital data or information, as messages or code. In this informational space, biological components acquire a new flexibility and mobility, since they can be distinguished from the biological substrate, the organisms, in which they are located. For all practical purposes, their ties to specific living organisms can be severed, so that they can be decoded and recoded, translated from "wet" physical samples of DNA to "dry" information, that can be stored, transported, and reassembled: they are interchangeable. Molecularization, writes Rose,

> strips tissues, proteins, molecules, and drugs of their specific affinities – to a disease, to an organ, to an individual, to a species – and enables them to be regarded, in many respects, as manipulable and transferable elements or units, which can be delocalized – moved from place to place, from organism to organism, from disease to disease, from person to person. Whether it is the transfer of genes from one species to another, the transfer of treatments from one disease to another, or the transfer of tissues, blood plasma, kidneys, stem cells, molecularization is conferring a new mobility of the elements of life. (2007: 15)

In this setting, the organic unity advanced by the molar model is dismantled, fragmented into interchangeable parts – not only on the molar level (hips, corneas, hearts and kidneys) – but also on the molecular level where vital elements can be re-engineered by molecular manipulation, customized and cloned.

At the level of genetic code, the molecular reconfiguration of living organisms erases essential differences between bodies – of the same species, and of different species (and to a certain extent, of species themselves, as we shall see shortly). This apparent indifference to species distinction (Waldby 2000) is what allows a good part of genetics research to advance, from comparing the genomes of humans to those of mice, rats, worms and even bacteria, to using non-human models such as the fruit fly to carry out experiments and manipulations that cannot be done on humans. The analogous and mobile nature conferred to genetic material by molecular biology has also opened up immense possibilities for transgenic work, where genetic material is exchanged between humans and animals. The ability to recombine genes from different species has enabled mice and goats to express human genes, plants to express genes from fish, and sheep to produce human proteins. While "cybrids", in which a human nucleus is implanted into an animal cell, and "chimeras", in which human cells are mixed with animal embryos, are proposed as new ways of obtaining personalized human embryonic stem cells. If skepticism often prevails regarding the usefulness of these experiments, they nevertheless undoubtedly indicate a novel correlation between human and animal bodies, as milieus through which analogous and transferable genetic information flows.

Thus the shift from the clinical to the molecular gaze does not consist in the mere extension of the reach of the clinical gaze, to regions that escape visualization by the naked eye, but in a reconfiguration of life processes and the body. While the molecular gaze continues to hold the body as an object of scrutiny, it is not so much as a functional organic whole but as an assemblage of molecular entities that can be identified, isolated and more than ever mobilized and manipulated. At the molecular level, vital elements and mechanisms have been deterritorialized from their previous settings of bodies, organs and organisms, a development that challenges the bounded nature of these entities. The molecular body is a fragmented, analogue body, made up of ever more precise but ever more transferable biological elements. The emphasis here is less on the body's essence as a functional organic whole than on its correlation with other bodies – less on what it *is* than on what it *does*.

5.1.3   The Body as Assemblage

The distinction between the molar and the molecular plays a central role in the thought of Gilles Deleuze and Félix Guattari (1987), insofar as they attempt to bring to light the wealth of molecular configurations that are often obscured and contained by the transcendental illusions of molarity. They offer one of the most comprehensible accounts of the molecular body that it is worthwhile to look at in some detail, especially insofar as the work of Deleuze and Guattari is a major influence among a number of radical posthumanists. For Deleuze and Guattari, molar configurations are identifiable wholes – such as an organism or a society – in which the connections have become stabilized and homogenized, while molecular configurations are assemblages in which elements are correlated in a non-rigidified manner and the boundaries of which are constantly fluctuating. Just as that the difference between the molar and the molecular body is not merely a question of the scale at which it is conceived, for Deleuze and Guattari the difference between molarity and molecularity is not one of size or scale but of composition and organization.

In the model of the molar body, organs and energies find their meaning and function through their integration into a coherent whole – the overarching structure of the unified body. For Deleuze and Guattari, bodies are neither unified, coherent, nor static. They are collections of disparate flows, organs, materials, energies and intensities, that congeal under particular conditions in relations with flows and intensities of surrounding objects to produce transitory but functional assemblages. Bodies are concoctions of material components, chemical compounds and social practices that, while giving rise to functional units, are not subordinated to an overriding organization or order.[4] Deleuze originally developed this notion of the body following Spinoza (Deleuze 1988), for whom the body is neither a locus of consciousness nor an organically determined entity. With Spinoza, as Deleuze and Guattari read him, the body is a multiplicity, a nexus of interconnections that is open to its surroundings and that can be composed, decomposed and recomposed by other bodies. This is because living beings are not grasped in terms of their form and function, rather, they are defined (and distinguished from other things) *kinetically* – in terms of relations of speed (motion and rest), and *dynamically* – in terms of their capacities for affecting and being affected. For Deleuze and Guattari, these two conditions, or axes, the kinetic and the dynamic, form a "social cartography" along which individuals can be mapped:

> A body is not defined by the form that determines it nor as a determinate substance or subject nor by the organs it possesses or the functions it fulfills. On the plane of consistency, *a body is defined only by a longitude and a latitude*: in other words the sum total of the material elements belonging to it under given relations of movement and rest, speed and slowness (longitude); the sum total of the intensive affects it is capable of at a given power or degree of potential (latitude). (Deleuze and Guattari 1987: 260, emphasis in original)

Following Spinoza, the body is understood more in terms of what it can *do* than in terms of what it *is*: "We know nothing about a body", they write, "until we know what it can do" (1987: 257). The pivotal term in this understanding of the body is "affect", the capacity a body has to form specific relations: what a body can do is determined by what its capacities to affect and be affected are (its affects), what it can perform, the linkages it establishes, the transformations it undergoes, the connections it forms with other bodies. Defining bodies in terms of the affects of which they are capable is equivalent to defining them in terms of the relations into which they can enter with other bodies, or in terms of their capacities for engagement with the powers of other bodies. In terms of affect, bodies undergo modification or change when they act upon other bodies or when they are acted upon by other bodies, in which different extensive relations and new intensive capacities can emerge.[5] This is not a functionalist approach; it rejects efforts to define the essential nature of a body. Rather, it consists in counting the affects of a body, its psychological, emotional and physical attachments. What characterizes bodies and distinguishes them

---

[4] This type of thought on the body also develops from extracting the notion of the body from its purely human and even biological setting. The body as assemblage holds not only for animal bodies, but for all bodies, be they chemical, political, biological or social: "a body can be anything; it can be an animal, a body of sounds, a mind or an idea; it can be a linguistic corpus, a social body, a collectivity" (Deleuze 1988: 127).

[5] What has been called a "turn to affect" occurred in critical and cultural theory in the mid-1990s, following the work of Deleuze and Guattari. This was often a response to the limitations of poststructuralist thought and deconstruction, namely, the problems associated with writing the body out of theory and the insistence on social structures rather than interpersonal relationships as formative of the subject. Contributions to this approach include Brian Massumi (1996, 2002), Eve Sedgwick (2003) and Patricia T. Clough, who edited *The Affective Turn in Social Theory* (2007). In this approach, affect is usually conceptualized as pre-individual bodily forces augmenting or diminishing a body's capacity to act.

from other bodies is not some a priori essence, but the speed and slowness, motion and rest, of the parts which compose them. Asking what a body can do locates the body as an active, experimenting, engaged and engaging body, and identifies Spinoza's ethics as an ethology, "the study of the relations of speed and slowness, of the capacities for affecting and being affected that characterizes each thing" (Deleuze 1988: 125).

Ian Buchanan, in light of this, has argued that the problem of what a body can do is best understood as an attempt to replace "aetiology (cause and effect) with ethology (action and affect), Freud with Spinoza" (1997: 74). The move from aetiology, the search for the cause of disease, to ethology, Buchanan explains, involves a change of direction: where aetiology looks backwards, ethology looks forward, and where aetiology looks inwards, ethology looks outwards. Ethology highlights similarities and differences in terms of a body's powers of affecting and being affected, rather than through taxonomic categorizations (the recognition, for example, that a draft horse has more in common with an ox than with a race horse).[6] The stuff of bodies, then, is the actual linkages between things. "Every relationship of forces", Deleuze writes, "constitutes a body – whether it is chemical, biological, social or political" (1983: 40). Bodies are not the locus at which forces act, they are the production of the interactions of forces.[7]

For Spinoza, in Deleuze's reading, the affects of the body correspond to the transition of the affected body from one state to another, and he distinguishes between those transitions that increase a body's power of acting – which give rise to joy – and those transitions that decrease a body's power of acting – that give rise to sadness (Deleuze 1988: 49–50). This point is of particular importance to the discussion on the celebratory nature of radical posthumanism's engagement with new technologies. In the distinction between joy and sadness posited by Spinoza and taken up by Deleuze and Guattari, there is something essentially *positive* in extroversion and something essentially *negative* in introversion. Moira Gatens and Genevieve Lloyd write that,

> there is in this contrast an inherent orientation of joy towards engagement with what lies beyond the self, ... and there is a corresponding orientation of sadness towards disengagement and isolation. The force of desire arising from joy will be strengthened, rather than weakened, by the power of external causes (Gatens and Lloyd 1999: 53).

The openness of the molecular body to connections and relations is seen as a very positive quality, and the opening up of the body to ever more connections is an ethical practice. For Deleuze, a body must strive to increase its capacity to be affected. Indeed, the notion of health in this framework is no longer considered as the balance between coordinated "mechanical" systems of the organism as

---

[6] This is one of the ethological relationships Deleuze and Guattari mention. They also argue that children often think of bodies in terms of affect (making them Spinozists by nature). Thus, Freud's Little Hans thinks of the horse in terms of affects such as "having eyes blocked by blinders, having a bit and a bridle, being proud, having a big peepee-maker [and] pulling heavy loads" (1987: 257).

[7] In this sense, Deleuze and Guattari are naturally indebted to Nietzsche, for whom the body is a composition of forces and should be understood in terms of quantities and qualities of forces.

advocated by modern biomedicine, rather it is directly related to the body's capacity to be affected. Thus relations that lead to the formation of new assemblages are considered healthy while those that disrupt old assemblages without forming new ones are unhealthy. "Health", explains Buchanan, "is the happy union of a capacity to form new relations and the new relations themselves, which in their turn permit the body to go on to form other new relations" (1997: 82).[8]

5.2   From Molar to Molecular Organisms

As we have seen, molar corporeality configures the body as a unified organic whole, a bounded, autonomous, self-regulating *organism*. The discussion on the molecular body has been articulated until now mostly in terms of human or even animal bodies. But organisms abound at all biological levels and have played a central role in biology since its birth as a separate and distinct science in the beginning of the nineteenth century. A strong analogy then, not surprisingly, can be drawn between the molar body and the modern biological organism, and in light of this analogy it becomes possible to detect the shift from the molar to the molecular configuration of the body – reconstrued as the organism – in some recent work in evolutionary theory as well. In the framework of evolutionary biology, we can understand an organism as molar firstly insofar as it replicates and reproduces itself without the intervention of what are commonly defined as "external" factors, that is, its traits are "passed down" in a continuous passage of vertical descent. And secondly, insofar as it is a bounded entity, separate from its environment and capable of self-regulation and self-formation.

In the past several decades, various new approaches to the relationship between organisms and their environments, backed by models of dynamical biological systems, and the discovery of non-vertical means of transferring genetic material (what is known as horizontal gene transfer and endosymbiosis) have begun to present new challenges to the molar understanding of biological organisms. If the molecular model of the body in the life sciences pertains mainly to what a body can do, insofar as biological elements can be isolated, decoded and recombined, these new findings in evolutionary biology indicate an a priori molecularity, insofar as organisms are the result of an active interaction with their environment and insofar as the origin of a significant number of organisms are located in processes of hybridization and inter-species genetic exchange. In other words, even more than the molecularization

---

[8] See for example, Deleuze and Guattari's analysis of anorexia (1987). Following Spinoza, they understand appetite or hunger not as a primary, inbuilt instinct, but as the product of a relation with food. The anorexic body is thus seen as incapable of realizing this productive relation with food, and anorexia is interpreted as an attempt to liberate the body from the insupportable burden of automatic relations. In this sense, anorexia is not a clinical condition but a practice of self, an attempt to produce a body without organs (Deleuze and Guattari's opposition to traditional medical interpretations of anorexia is also related to its psychologist aspect, that explains anorexia as a body-image disorder. This kind of reading sees the body as subordinate to the mind – a view they reject entirely).

of the body in the life sciences, the presence of a similar shift in evolutionary theory makes a strong case for the presence of a non-humanist originary biological prostheticity trend in biology.

5.2.1   Evolutionary Biology: The Organism as Dynamical System

Dynamic and complex models of living systems have recently been adopted by a number of leading biologists to emphasize the co-evolution of the organism and its environment. Molarity is challenged here to the extent that the opposition between organism and environment is undermined.[9] This opposition has played a fundamental role in the history of modern biology. Indeed, it was not always the case, since before Darwin and Pasteur, biologists like Lamarck argued for a continuity between environment and organism. In her study on the relationship between genes and the organism over the last century, Evelyn Fox Keller (2000) argues that the attempt to answer the question "What is an organism?" can be taken as marking the beginnings of biology, that science that separated the world into the living and the nonliving (2000: 106). It was Kant, Keller explains, who offered one of the first modern definitions of an organism in his *Critique of Judgment*, as "*an organized natural product … in which every part is reciprocally both end and means*. In such a product nothing is in vain, without an end, or to be ascribed to a blind mechanism of nature" (1993: 66, p. 558). For Kant, then, it was the internal dynamics, the capacity for self-organization, i.e., the absence of an external organizing force, that characterized the living organism.[10] The modern organism was thus coined as a bounded, autonomous physiochemical entity that was capable of self-regulation and self-generation by virtue of its organization.

Models of dynamical processes argue for an undoing of the opposition erected between the organism and the environment, insofar as an environment cannot be separated from what organisms are and what they do (Kampis 1991; Kauffman 1993; Goodwin 1995). Here both the reification of the role of DNA and the reification of the environment as that which selects organisms are challenged. Organisms, it is argued, cannot be treated as closed systems that evolve separately from their environment and are subjected to external forces in a passive model of adaptation. Rather, they should be understood as active participants that interact with their environments and "select" them as well.

---

[9] These models need to be understood as an attempt to go beyond the alleged genetic reductionism and determinism of neo-Darwinism rather than as an opposition to classical Darwinism. Though, if neo-Darwinism views genes rather than organisms as the irreducible and basic elements of biological reality, this gene-centrism is in fact derived from Darwin's organism-centrism. I will return to a discussion on genetic reductionism in much greater detail in Chap. 8.

[10] For Kant, this emphasis on *self*-organization serves as both an opposition to argument from design and as a damper on the fascination with the very lifelike automata of his time.

Brian Goodwin (1995), for example, who advocates a union of complexity theory and evolutionary biology, argues that as an organism matures, characteristic types of order emerge from a chaotic interaction of genes, molecules and the environment. What we see as highly ordered features of the development of organisms should not be regarded as the accomplishment of natural selection, but of the innovative capacities of self-organization of complex genetic regulatory systems. Here, cooperation and webs of relationships are seen as playing a role as important as competition and inheritance. Variation on this model is thus not simply the result of random genetic mutation but of "the intrinsically regulative and plastic responses of the organism to its environment during its life-cycle" (Goodwin 1995: 104–5). Organisms, such biologists argue, develop internal structures which serve to mediate the environment, and they should be understood in more dynamical terms than the rigors of neo-Darwinism allows, as *open* systems that exchange energy or matter and information with their environment. In this co-evolutionary view the relationship between organisms and environment is seen to rest on a series of feedback loops.

Developmental systems theory, advocated by theorists like Susan Oyama (2000a, b), also attempts to bring this relationship into focus by offering an alternative view of what heredity is, as the construction of developmental means rather than the transmission of genetic information. Proponents of developmental systems argue that developmental means also include the complex machinery of the cell, the maternal reproductive system, the care of parents, and interdependent relations with many aspects of the environment. Developmental systems are the changing complex of interacting influences, some of which are inside the organism and some of which are outside, that contribute to form and variation. Susan Oyama contends that the dichotomous construction between predisposing genes and accidental development reinstates the misleading nature/nurture opposition and that in biology this construction is predicated on the belief that information can preexist the processes that give rise to it (2000b: 15–16). Oyama argues that the term information here functions within the much older philosophical discourse of the distinction between form and matter, by assigning formative relevance only to the gene and by reducing the transmission of form or meaning to the organism. Rather than an idealization of the gene as the code of information that determines life, Oyama views the creation of information through the reciprocal selection and joint co-action between genes and environments (2000b: 33). Biological information is created in interaction.

The molecular biologist and feminist theorist Anne Fausto-Sterling has expanded developmental systems theory beyond strictly biological understandings in her work *Sexing the Body: Gender Politics and the Construction of Sexuality* (2000). She claims that developmental systems theory erodes the nature/nurture distinction, and that the opposition between nature and culture as contenders in shaping both the body and subjectivity should be replaced by a system of dynamical processes and relations between internal and external environments, sociocultural and biological/material factors. While her research focuses on the construction of sexuality and categories of difference, her notion of a "biocultural system" in which cells and culture are mutually constituted holds for all of biology. Cultural or discursive experiences, she argues, are not only inscribed on the surface of bodies, but go literally beneath the skin as "events outside of the body become incorporated into our very flesh" (2000: 238).

On her model, socioculturally-shaped behavioral patterns as well as reactions of the neural system to external signals affect one's muscles, bones, nerves and even the architecture of one's cells, so that cells are continuously being formed and reformed through their relations with their internal and external environment.

The view of the molar organism has also been recently undermined by epigenetic research (Jablonka and Lamb1999; Jablonka 2004), the study of changes in gene activity that do not involve alterations to the genetic code but still get passed down to at least one successive generation. While biologists have been aware of epigenetic marks since at least the 1970s, until recently epigenetic phenomena were regarded as relatively insignificant in the greater scheme of DNA-dominant models of development and evolution. But biologists are finding that environmental conditions, such as diet or stress, can leave an imprint on the genetic material of eggs and sperm – that non-genetic variation acquired during the life of an organism can be passed on to offspring. Thus an increasing number of studies are reporting evidence that links the environment to long-lasting effects on phenotype: when fruit flies are exposed to certain chemicals, at least 13 generations of their descendants are born with bristly outgrowth on their eyes; exposing a pregnant rat to a chemical that alters reproductive hormones leads to generations of sick offspring; the children and grandchildren of individuals who were malnourished in adolescence show higher rates of heart disease and diabetes; and epigenetic factors may explain differences in disease susceptibility among identical twins.

The source of variation in subsequent generations in each of these cases is not DNA, and shows that "epigenetic inheritance is ubiquitous", state Eva Jablonka and Gal Raz (2009: 131), in a review that catalogs some 100 forms of epigenetic inheritance in bacteria, protists, fungi, plants, and animals. As these authors claim, epigentics poses a serious challenge to the "Modern Synthesis" version of evolution which states that variations are random, genetic, and very gradual, and the incorporation of the epigenetic perspective into evolutionary theory heralds a "new extended theory" that would also be informed by Lamarckian frameworks and developmental studies (Jablonka and Raz 2009: 168). But epigenetics also has important repercussions for how we understand bodies and organisms. Health and development in the epigenetic perspective are a result of a combination of the individual genotype and environmental pressures, which interact in more intimate ways than allowed for in the molar configuration, where inside and outside are clearly demarcated.

5.2.2   Molecular Phylogeny: From the Tree to the Network of Life

Non-Vertical Means of Gene Transfer

The dynamical models reviewed here challenge the view that the development of the organism follows its course quite independently of a predominantly passive "external" environment, be this at the general level of the organism or the more basic level of genes. Heredity, in this molar configuration of the organism, takes place via the transmission of genes from generation to generation, in a linear pattern. Epigenetics and developmental studies show that this is not always the case, insofar as information also "comes in" from the outer environment, contributing in important ways to what an organism is. In another field of biology, molecular phylogeny – the analysis of evolutionary relationships – this view is running up against other problems, whereby the linear transmission of genetic information *within* species is questioned by evidence of genetic transfer *between* species.

The assumption that traits are only passed down vertically has been a central tenet of evolutionary biology at least since Darwin. It is the vertical and direct transmission of traits from parent to progeny that is the basis for the evolutionary relatedness of Darwin's famous "Tree of Life" schema, which illustrates the evolutionary relationships between species as a vast, ever-bifurcating structure that grows out of the Last Universal Common Ancestor at its base and splits into an increasing number of branches. This model of a single genealogical tree has been the unifying principle for understanding the history of life ever since. But it was not until recently that scientists could do much more than infer the relatedness of organisms by comparing their anatomy or physiology, and this almost only for complex organisms, not for the microscopic single-celled organisms that have been the main inhabitants of the planet for at least half of its existence. In the 1960s, prompted by new techniques of DNA, RNA and protein sequencing developed in molecular biology, the new field of molecular phylogeny could begin to compare sequence analyses from different species (Durbin et al. 1998, Hall 2004). This led to a great accumulation of data from sequenced genomes, which, it was believed, would finally confirm the molecular certainty of the tree of life model, simply by demonstrating that the closer two species were located on the tree the closer their genomes would prove to be, or by establishing additional branching patterns to the original schema.[11]

But the genetic information derived from DNA sequencing has also shed light on phenomena that challenge the standard representations of species lineage and show that genetic material does not always follow a strict vertical pattern. The amassment of sequencing data in the 1970s and 1980s soon revealed that many patterns of relatedness between the bacteria and archaea domains of life indicated that genes were being routinely transferred horizontally, *across* taxonomic domains, and not only vertically, within them. The process of what is known as horizontal, or lateral gene transfer (LGT), at first thought to be a minor factor in the transfer of genetic information, soon proved to be a central feature of the evolutionary of cells, which have been found to possess a significant amount of bacterial genes (Bapteste et al. 2004;

---

[11] These techniques did indeed lead to the unexpected discovery of an important new branch on the tree of life. Before this time it was generally believed that the world of living things could be divided into two separate groups depending on cell-structure: bacteria, or prokaryotes (organisms composed of cells with no nucleus) and eukaryotes (organisms composed of cells that contain a nucleus, such as animals, plants, fungi and many unicellular life-forms). In the late 1970s, thanks to the new genome sequencing techniques, a new group made up of unicellular archaea – prokaryotes previously believed to be bacteria but found to have a radically different molecular structure – was revealed, leading to a new three-domain view of life. The archaea came to be seen as an intermediate domain of life between bacteria and eukaryotes.

Garrett et al. 2007).[12] Such transfer involves the delivery of single genes, or whole suites of them, not from a parent cell to its offspring, but across the species barrier. Furthermore, such "gene swapping" revealed itself not to be limited to crossings between bacteria and archaea but also to be widespread in the third domain of life, the eukaryotes, those complex organisms that are the building blocks of all large forms of life, via a process known as endosymbiosis.

Endosymbiosis refers to the process by which an independent organism is incorporated by another independent organism and the two gradually fuse over time. The biologist Lynn Margulis (1970) first put forward the endosymbiont hypothesis as an alternative explanation for how eukaryotes evolved from prokaryotes.[13] According to the conventional tree of life model, the eukaryotes evolved from prokaryotic ancestors that accumulated mutations over time, until gradually arriving at the structural level of eukaryotic cells. Instead, Margulis proposed that the origin of eukaryotic cells can be traced to symbiotic relationships developing inside one prokaryotic cell (a host) with other prokaryotes (guests) that it incorporated. Such guest cells might have gained entry into host cells as undigested prey or as internal parasites, following which the combination became mutually beneficial to both host and guest organisms, rendering them increasingly interdependent, and leading to the emergence of novel metabolic capabilities in at least one of the partners. There is a substantial amount of evidence to suggest that this theory of the evolution from prokaryotes to eukaryotes is correct, at least with regard to how eukaryotic cells first came to possess mitochondria and chloropolasts.

Increasing awareness of processes of LGT and endosymbiosis and the extensive evidence that these have left an indelible mark on all organisms, defies the concept of a hierarchical universal classification, and challenges the molar view of the organism whose replication is based on the vertical inheritance of traits intra-species. And if unrelated organisms are swapping genes back and forth, then the model of a Tree of Life cannot capture much of what is important in the evolutionary process.

A Web of Viral Life

For a number of biologists and evolutionists today, processes of horizontal gene transfer and endosymbiosis has led to a questioning of the accuracy of the Tree of Life hypothesis, indicating that the tree image is oversimplified and that alternative models, or "pattern pluralism", would better explain the history of evolving forms (Doolittle 2000; Doolittle and Bapteste 2007; Dagan 2006). Namely, incorporating these processes into the bigger picture of evolution would mean that the base of the

---

[12] Tal Dagan (2008) and colleagues at the Heinrich Heine University in Düsseldorf found that more than 80 % of genes in genomes from some 181 prokaryotes were involved in horizontal gene transfer.

[13] The idea that the eukaryotic cell is actually a colony of microbes was first suggested in the 1920s by the American biologist Ivan Wallin (Fausto-Sterling 1993). Margulis is the originator of the modern version of endosymbiosis. See also her *Symbiosis in Cell Evolution* (1981) and, in collaboration with Dorion Sagan, *Microcosmos: Four Billion Years of Evolution* (1986).

tree looks much more like a web, or a bush, than a trunk. This revised picture undermines the notion of a single common ancestor, the idea that one single organismal lineage lies at the origin of all of life.[14] For many evolutionary biologists today this is not to say that the Tree of Life needs to be "uprooted" or completely chopped down. This is because non-vertical gene transfer is characteristic of the earliest stages of evolution, when all organisms were single cells, a lot more than of later stages, where multicellular animals are involved. Such a net, or web, of life would be illustrative of evolution *before* the distinction between eukaryotes and prokaryotes was discernible – once the main three groups rose out of the web, their evolution can be depicted as branching out vertically, so that at the "top" a tree model would be appropriate for multicellular animals, plants and fungi.[15]

But for some biologists, LGT and endosymbiosis entail a need for fundamental reform of our understandings of evolution, and the development of a more pluralistic description of the history of biodiversity (O'Malley and Dupré 2007a; Rose and Oakley 2007). Eric Bapteste and Richard Burian (2010), for example, argue that the ubiquity of LGT in the natural world requires a type of "lateral thinking" in evolutionary patterns, that could be generated with network representations. Marc Ereshefsky (2010) argues that because the majority of life now and throughout evolutionary history has been microbial, no universal concept can cover all of life. Species, in this sense, exist only as pragmatically defined categories and not as anything essentially "real". It is therefore important that we become aware that the algorithms used to study evolutionary hierarchies often impose or extract a single tree model on a highly complex natural world.

Outside of biology strictly speaking, Deleuze and Guattari (1987) offer the image of the "rhizome" as just such an alternative to the Tree of Life as a metaphor for how we think. A rhizome is a continuously growing underground stem, which develops

---

[14] Though Darwin himself concluded in *The Origin of Species* that all life arose from "a *few* forms or... one".

[15] In 2009, for example, many lay-readers and biologists were quick to denounce an article in the New Scientist entitled "Uprooting Darwin's Tree" (Lawton 2009), and its sensationalist cover, which read "Darwin was Wrong". This included a letter that appeared in the February 18 issue by Daniel Dennett, Jerry Coyne, Richard Dawkins and Paul Myers entitled "Darwin was right". Such criticisms usually made the point that these processes are only typical of single-cell organisms, and that in light of the fact that Darwin knew nothing of microorganisms or molecular genetics, his tree is quite accurate. But what's more, they often expressed outrage at the claim that "Darwin was wrong", which they believed would play into the hands of creationists. As they feared, shortly after its publication members of the board of education of the state of Texas were already citing the article as an indication that creationist-inspired theories should be used in schools. Not that the staff at *New Scientist* did not anticipate this: as the editor wrote in the issue's editorial,

> None of this should give succour to creationists, whose blinkered universe is doubtless already buzzing with the news that "*New Scientist* has announced Darwin was wrong". Expect to find excerpts ripped out of context and presented as evidence that biologists are deserting the theory of evolution en masse.

The controversy surrounding the article is a telling indication of how intense the clash between evolutionary biologists and creationists today really is. Nonetheless – granted the article, and especially the cover, were sensationalist – it would be very unfortunate if a fear of legitimizing creationism were to become the source of a Darwinian orthodoxy.

by producing adventitious lateral shoots, a network of multiple branching roots. Unlike the arborescent figure, they argue, that develops genealogically and teleologically through filiation and descent, the rhizome has no unified point of origin (it develops from the middle), no central axis and no given direction of growth; it operates via "variation, expansions, conquest, capture, offshoots" (1987: 21), and any of its points can be connected to any other.[16]

Furthermore, the general argument against phylogenetic trees and the need to rethink our models of evolution for those biologists discussed here, does not just ensue from the fact that single-cell organisms make up at least 90 % of all known species. Frédéric Bouchard (2010) argues for a recognition of the more radical implications of LGT for biological lineages and individuals. It won't do, he claims, to restrict LGT to the microbial world, that "blackboxes" a process that occurs in the *macrobial* world as well. Bouchard explores the superorganismal world and how considerations of symbiotic associations of organisms destabilize the spatial and temporal boundaries of organisms and lineages. Hybridization, for example, the fusion of two separate lineages, has been found to play a role in plant and animal species, and there is increasing evidence of cases of LGT from bacteria, viruses and even other animals to multicellular organisms, including insects, fish and cows.[17] Endosymbiosis has also been found to occur at "higher" levels, including plants. If such blurring of species boundaries and genetic flux across domains of life are much more common in microbes than in plants and animals, these findings do demonstrate that they are evolutionarily important.

For some biologists this suggests that evolution may be much more about mergers and collaboration than vertical descent. For Margulis (1986), for example, symbiogenesis is the engine of biodiversity and a driving force behind evolution. On her model, cooperation, interaction and mutual dependence among life forms are what allowed for life's eventual global dominance. Rather than focus on the elimination of competitors, Margulis' view of evolution downplays competition itself on the basis of symbiotic relationships.[18] Genetic variation here is not the outcome of a

---

[16] Many theorists have drawn a parallel between Margulis' theorization of endosymbiosis and Deleuze and Guattari. See especially Keith Ansell Pearson's *Viroid Life* (1997) and *Germinal Life* (1999), Manuel DeLanda's *Intensive Science and Virtual Philosophy* (2002) and *A New Philosophy of Society* (2006). See also Rosi Braidotti (2002, 2006a), Jon Protevi (2006) and Mark Hansen (2000).

[17] Some of these examples include the cow genome, which was found to contain a piece of snake DNA that transferred horizontally some 50 million years ago, the genome of a fruit fly, which contains the entire integrated genome of the bacterium *Wolbachia*, and a gene crucial to the function of stinging cells in jellyfish and sea anemones found to be transferred from bacteria.

[18] Margulis has been accused of over-emphasizing these cooperative aspects of evolution over competition, and of inferring more general normative conclusions from here. See, for example, the comments made by leading scientists about her in *The Third Culture* (Brockman 1995: 140–141, 145), such as Daniel Dennett: "I think she's trying to take a wonderful idea and harness it as a political idea", George C. Williams: "Margulis is very much afflicted with a kind of 'God-is-good' syndrome", and Francisco Varela: "It's unfortunate that she has veered into some weird second stage". This has to do mostly with Margulis' collaboration with James Lovelock in the development of the "Gaia hypothesis". On this see Doolittle's article "Is Nature Really Motherly?" (1981), in which he argues that there is nothing in the genome of individual organisms that can provide the feedback mechanisms the Gaia theory proposes, and Dawkins' *The Extended Phenotype* (1999),

gradual accumulation of mutations, but of bacteria-like reproduction, of contact or contagion that cuts across taxonomic borders.

Contagion is not just meant metaphorically here. It seems that viruses, thanks to their ability to copy DNA from one genome to another, are the main agents of this genetic flux across all domains of life. The era of genomics has revealed a rich picture of viruses as a creative evolutionary force, life's most fertile breeding ground for novel DNA sequences. Indeed, geneticists have found the remains of ancient viral infections in the genomes of all living organisms, not only bacteria. Retroviruses – viruses that convert their genome into DNA once they have infected a cell and integrate it into the host – have been found to be an important part of eukaryotic DNA, for example. And endogenous retroviruses (retroviruses that become a permanent addition of the host cell) have been found in many animals and humans (Forterre 2006). As Margulis and others (O'Malley and Dupré 2007b) have noted, the emerging view of life as viral challenges the idea that evolution is driven by the competition between "selfish genes" (Dawkins 1989), by highlighting microbial capacities for cooperation and communication. This co-evolution between organisms and microbes also dissolves the notion of the molar organism, which can no longer be seen as a distinct package of genetic information that has been passed along an unbroken line of ancestors, and suggests that the model of an interconnected network of circulating genes may be more appropriate. An endosymbiotic model of the body, for example, rather than viewing the body as a molar unity that is under attack by foreign agents, views many disease agents as normally present in the human body. Here health is more a matter of maintaining an ecology than defending a unity (Sagan 1992).[19] As Keith Ansell Pearson suggests in *Viroid Life* (1997), the model of endosymbiosis has a "filthy lesson to teach us: one of the human as an integrated colony of amoebid beings" (124).

5.3   Conclusion

In this chapter, I have attempted to show how the molar formulation of the body or the organism, which is another articulation of the humanist dualist paradigm on the corporeal level, is not only being challenged in the realm of technology but also in

---

where he argues that the type of working in concert necessitated by the Gaia hypothesis requires of organisms a foresight that they do not have.

[19] Sagan cites the streptococcus bacteria and *Candida albicans* fungi. This approach is also becoming widespread in cancer immunobiology. Here the idea is that cancer cells are present in all bodies, but that the immune system usually manages to keep these early cancers and pin-headed tumors in check. See "The Immunobiology of Cancer Immunosurveillance and Immunoediting" by Dunn et al. (2004)), and "Cancer without Disease" by Judah Folkman and Raghu Kalluri (2004). Folkman and Kalluri cite autopsy studies which have revealed that more than a third of women aged 40–50 have small in situ breast carcinomas, whereas only 1 % are diagnosed with clinical breast cancer, analogous findings that hold for prostate cancer in men, and autopsies that show that virtually all people aged 50–70 have small in situ thyroid tumors, yet well below 1 % are diagnosed with clinical thyroid cancer.

biological research. Lateral gene transfer and endosymbiosis, epigenetics and developmental theory, transgenics and hybridization, are processes that function *through* organisms; within and without and among them. They challenge the configuration of organisms as bounded, functional units, as molar wholes, and offer molecular models of organisms as fragmented and open-ended assemblages: that can be relocated and re-engineered, in the context of molecular biomedicine, exchanged and negotiated with the environment, in the context of developmental theory and epigenetics, or swapped across vast taxonomic breadths, in the context of molecular phylogeny.

The discussion on molecular organisms and the evolutionary history of microbes may seem far removed from the exploration of the implications of emerging biotechnologies for what it means to be human. But what we find here is that notions like networks, mediation and originary prostheticity appear to not only be a feature of the relationship between humans and technology but also of the relationship between organisms, species and domains of life, that have typically been considered distinct and independent. The shift from molar to molecular bodies in these models can thus be seen as a posthumanist movement that mirrors the one undertaken by methodological and radical posthumanism. Like methodological and radical posthumanism, these biological models attempt to move beyond essentialism, insofar as their emphasis is often on mechanisms rather than units of heredity, on what organisms do rather than what they are, and beyond dualism, by challenging the nature/nurture, subject/object and inner/outer distinctions. Like methodological and radical posthumanism, these models can be seen as exploring the claim for the interwoven nature of the human with its material environment – and of offering a very serious answer to the question posed in the "Cyborg Manifesto" as to "Why should our bodies end at the skin?" (1991: 178). What's more models of corporeality and models of subjectivity often reflect and reproduce each other. And if the molar body-as-organism is being displaced and reformulated in terms of complexity, open systems and molecular mobility, then the molar subject as an autonomous, bounded, unitary self, surely will not remain intact. This assumption leads us to the next chapter and the discussion on posthuman subjectivity, or what kind of non-humanist subjectivity is implied by molecular corporeality.

References

Abir-Am, P. (1985). Themes, genres, and orders of legitimation in the consolidation of new scientific disciplines. *History of Science, 23*(59), 73–117.

Ansell Pearson, K. (1997). *Viroid life: Perspectives on Nietzsche and the transhuman condition.* London/New York: Routledge.

Ansell Pearson, K. (1999). *Germinal life: The difference and repetition of Deleuze and Guattari.* London: Routledge.

Bapteste, E., & Burian, R. M. (2010). On the need for integrative phylogenomics, and some steps toward its creation. *Biology and Philosophy, 25*(4), 711–736.

Bapteste, E., Boucher, Y., Leigh, J., & Doolittle, W. F. (2004). Phylogenetic reconstruction and lateral gene transfer. *Trends in Microbiology, 12*(9), 406–411.

Bouchard, F. (2010). Symbiosis, lateral function transfer and the (many) saplings of life. *Biology and Philosophy, 25*(4), 623–641.

Braidotti, R. (2002). *Metamorphoses: Towards a materialist theory of becoming*. Cambridge: Polity Press.

Braidotti, R. (2006). Affirming the affirmative: On nomadic affectivity. *Rhizomes, 11/12*(2005/2006). http://www.rhizomes.net/issue11/braidotti.html. Accessed 13 June 2013.

Brockman, J. (1995). *The third culture: Beyond the scientific revolution*. New York: Simon & Schuster.

Buchanan, I. (1997). The problem of the body in Deleuze and Guattari, or, what can a body do? *Body & Society, 3*(3), 73–91.

Canguilhem, G. (1994). The concept of life. In F. Delaporte (Ed.), *A vital rationalist: Selected writings from Georges Canguilhem* (pp. 303–320). New York: Zone.

Chadarevian, S., & Kamminga, H. (Eds.). (1998). *Molecularizing biology and medicine: New practices and alliances, 1910s-1970s*. Amsterdam: Harwood Academic Publishers.

Clough, P. T., Hardt, M., & Halley, J. (Eds.). (2007). *The affective turn: Theorizing the social*. Durham: Duke University Press.

Dagan, T., & Martin, W. (2006). The tree of one percent. *Genome Biology, 7*(10), 118.

Dagan, T., Artzy-Randrup, Y., & Martin, W. (2008). Modular networks and cumulative impact of lateral transfer in prokaryote genome evolution. *Proceedings of the National Academy of Sciences of the United States of America, 105*(29), 10039–10044.

Dawkins, R. (1989). *The selfish gene*. Oxford: Oxford University Press.

Dawkins, R. (1999). *The extended phenotype: The long reach of the gene*. Oxford: Oxford University Press.

DeLanda, M. (2002). *Intensive science and virtual philosophy*. London: Continuum.

DeLanda, M. (2006). *A new philosophy of society: Assemblage theory and social complexity*. London: Continuum.

Deleuze, G. (1983). *Nietzsche and philosophy* (trans: Tamlinson, H.). New York: Columbia University Press. Original edition, 1962.

Deleuze, G. (1988). *Spinoza: Practical philosophy* (trans: Hurley, R.). San Francisco: City Lights. Original edition, 1970.

Deleuze, G., & Guattari, F. (1987). *A Thousand Plateaus: Capitalism and schizophrenia* (trans: Massumi, B.). Minneapolis: University of Minnesota Press. Original edition, 1980.

Doolittle, W. F. (1981). Is nature really motherly? *CoEvolution Quarterly, 29*, 58–63.

Doolittle, W. F. (2000). Uprooting the tree of life. *Scientific American, 282*(February), 90–95.

Doolittle, W. F., & Bapteste, E. (2007). Pattern pluralism and the tree of life hypothesis. *Proceedings of the National Academy of Sciences, 104*(7), 2043–2049.

Dunn, G. P., Old, L. J., & Schreiber, R. D. (2004). The immunobiology of cancer immunosurveillance and immunoediting. *Immunity, 21*(2), 137–148.

Durbin, R., Eddy, S. R., Krogh, A., & Mitchison, G. (1998). *Biological sequence analysis*. Cambridge: Cambridge University.

Ereshefsky, M. (2010). Microbiology and the species problem. *Biology and Philosophy, 25*(4), 553–568.

Fausto-Sterling, A. (1993). Is nature really red in tooth and claw? *Discover, 14*(April), 24–27.

Fausto-Sterling, A. (2000). *Sexing the body: Gender politics and the construction of sexuality*. New York: Basic Books.

Folkman, J., & Kalluri, R. (2004). Cancer without disease. *Nature, 427*(6977), 787.

Forterre, P. (2006). The origin of viruses and their possible roles in major evolutionary transitions. *Virus Research, 117*(1), 5–16.

Foucault, M. (1973). *The birth of the clinic: An archaeology of medical perception* (trans: Sheridan Smith, A. M.). London: Tavistock Publications. Original edition, 1963.

Garrett, R. A., Klenk, H.-P., Walsh, D. A., Boudreau, M. E., Bapteste, E., & Doolittle, W. F. (2007). The root of the tree: Lateral gene transfer and the nature of domains. In R. A. Garrett & H.-P. Klenk (Eds.), *Archaea: Evolution, physiology, and molecular biology* (pp. 29–37). Oxford: Blackwell.

Gatens, M., & Lloyd, G. (1999). *Collective imaginings: Spinoza, past and present*. London/New York: Routledge.

Goodwin, B. (1995). *How the leopard changed its spots: The evolution of complexity*. London: Phoenix.

Gwyn, R. (2002). *Communicating health and illness*. London: Sage.

Hall, B. G. (2004). *Phylogenetic trees made easy: a how-to manual* (2nd ed.). Sunderland: Sinauer Associates.

Hansen, M. (2000). Becoming as creative involution?: Contextualizing Deleuze and Guattari's Biophilosophy. *Postmodern Culture, 11*(1). Available at http://muse.jhu.edu/journals/postmodern_culture/v011/11.1hansen.html. Accessed 21 August 2013.

Jablonka, E. (2004). Epigenetic epidemiology. *International Journal of Epidemiology, 33*(5), 929–935.

Jablonka, E., & Lamb, M. J. (1999). *Epigenetic inheritance and evolution: The Lamarckian dimension*. Oxford: Oxford University Press.

Jablonka, E., & Raz, G. (2009). Transgenerational epigenetic inheritance: Prevalence, mechanisms, and implications for the study of heredity and evolution. *The Quarterly Review of Biology, 84*(2), 131–176.

Kampis, G. (1991). *Self-modifying systems in biology and cognitive science*. Oxford: Pergamon Press.

Kant, I. (1993). *Critique of judgment* (trans: Meredith, J. C.). Chicago: Encyclopedia Britannica.

Kauffman, S. (1993). *The origins of order: Self-organization and selection in evolution*. Oxford: Oxford University Press.

Kay, L. E. (1993). *The molecular vision of life: Caltech, the Rockefeller foundation, and the rise of the new biology*. New York: Oxford University Press.

Kay, L. E. (2000). *Who wrote the book of life: A history of the genetic code*. Stanford: Stanford University Press.

Keller, E. F. (2000). *The century of the gene* (3rd ed.). Cambridge, MA: Harvard University Press.

Lawton, G. (2009). Uprooting Darwin's tree. *New Scientist, 24*(January), 34–39.

Margulis, L. (1970). *The origin of eukaryotic cells*. New Haven: Yale University Press.

Margulis, L. (1981). *Symbiosis in cell evolution*. San Francisco: W.H. Freeman.

Margulis, L., & Sagan, D. (1986). *Microcosmos: Four billion years of microbiological evolution*. Berkeley: University of California Press.

Massumi, B. (1996). The autonomy of affect. In P. Patton (Ed.), *Deleuze: A critical reader* (pp. 217–240). Oxford: Basil Blackwell.

Massumi, B. (2002). *Parables for the virtual: Movement, affect, sensation*. Durham: Duke University Press.

O'Malley, M. A., & Dupré, J. (2007a). Towards a philosophy of microbiology. *Studies in History and Philosophy of Science Part C: Studies in History and Philosophy of Biological and Biomedical Sciences, 38*(4), 775–779.

O'Malley, M. A., & Dupré, J. (2007b). Size doesn't matter: Towards a more inclusive philosophy of biology. *Biology & Philosophy, 22*(2), 155–191.

Oyama, S. (2000a). *Evolution's eye: A systems view of the biology-culture divide*. Durham: Duke University Press.

Oyama, S. (2000b). *The ontogeny of information: Developmental systems and evolution*. Durham: Duke University Press.

Protevi, J. (2006). Deleuze, Guattari and emergence. *Paragraph, 29*(2), 19–39.

Rabinow, P. (1992). Artificiality and enlightenment: From sociobiology to biosociality. In J. Crary & S. Kwinter (Eds.), *Incorporations* (pp. 234–252). New York: Zone.

Rose, N. (2001). The politics of life itself. *Theory, Culture and Society, 18*(6), 1–30.

Rose, N. (2007). *The Politics of life itself: Biomedicine, power, and subjectivity in the twenty-first century*. Princeton: Princeton University Press.

Rose, M. R., & Oakley, T. H. (2007). The new biology: Beyond the modern synthesis. *Biology Direct, 2*(30). doi: 10.1186/1745-6150-2-30.

Sagan, D. (1992). Metametazoa: Biology and multiplicity. In J. Crary & S. Kwinter (Eds.), *Incorporations* (pp. 362–385). New York: Zone Books.

Sedgwick, E. (2003). *Thinking feeling: Affect, pedagogy, performativity*. Durham: Duke University Press.

Sontag, S. (1977). *Illness as metaphor*. New York: Picador.

Waldby, C. (2000). *The visible human project: Informatic bodies and posthuman medicine*. London/New York: Routledge.

Chapter 6
Posthuman Subjectivity: Beyond Modern Metaphysics

Abstract This chapter explores the implications of posthumanist subjectivity via a discussion on subjectivity in the work of some important precursors of non-humanist posthumanism on subjectivity, such as Heidegger, Levinas and Deleuze, to methodological and radical posthumanists like Latour and Haraway. The human being is conceptualized here not as an independent and autonomous entity with clear cut boundaries but as a heterogeneous subject whose self-definition is continuously shifting, and that exists in a complex network of human and non-human agents and the technologies that mediate between them.

The discussion on subjectivity in the methodological and radical posthumanist approaches brings to light several significant shortcomings. Methodological posthumanism, after having argued for the agency of technological artifacts, too often fails to carry through the implications this has for human subjects. While radical posthumanism too often concedes to a celebration of hybridity (per se) and the claim that emerging biotechnologies have the potential to bring about a fundamental break with modernity. This critique serves as a platform to introduce the mediated posthumanist approach by reading Foucault's work on subject constitution via the notion of technological mediation and extending his notion of "technologies of the self" to biotechnologies. In this reading, the subject is constituted in specific ways by its technological mediations with the world, but it also develops an active relation to them, so that technologies can be seen as ethical practices that an interconnected, dynamic and molecular subject works with to constitute itself.

Keywords Posthuman subjectivity • Symmetry • Technologies of the self • Cyborg • Hospitality

Bodies are not outside of history. They are modes of organization which are both the effect of and productive of historically specific political, economic, and cultural formations. As the dominant model of corporeality in that political, economic and cultural formation called modernity, the molar body-as-organism has a privileged

T. Sharon, *Human Nature in an Age of Biotechnology: The Case for Mediated Posthumanism*, Philosophy of Engineering and Technology 14, DOI 10.1007/978-94-007-7554-1_6, © Springer Science+Business Media Dordrecht 2014

relationship to the dominant model of subjectivity of that same formation. Indeed, the attempt to think of these models of body and subjectivity separately is in itself problematic since they are conceptually co-dependent: modernist thought assumes that the body is the unambiguous locus of the self (agency, consciousness, mind), the ground of identity, a vessel occupied by and at the disposal of an animating, willful subjectivity. And it is specifically within a bounded, molar body-as-organism that the modern subject is located because it acts as the boundary, limit, edge or border of subjectivity, that which divides the subject both from other subjects and from objects in the world. "Modernity", writes Foucault in *The Order of Things*, "begins when the human being begins to exist within his organism, inside the shell of his head, inside the armature of his limbs, and in the whole structure of his physiology" (1989: 318). Thus, even as the model of modern subjectivity assumes a sharp dichotomy between self and body (a mind/body or form/matter dualism), it is the coupling of a single awareness of self and a single physical body, a harmonious unified cohesion of both, that gives rise to individual subjectivity in the modern sense.

The main implication of this corporeality/subjectivity bond is that any transformation in our dominant model of the body will or should effectuate a transformation in our dominant model of subjectivity, and vice versa. In the previous chapter I argued that the dominant model of molar corporeality is giving way, in some areas of research, to a less-bounded, increasingly fragmented and molecular model. The question that at present ensues is, is the molar model of the autonomous, fixed and bounded subject also undergoing a similar shift, and if so, what exactly does this molecular, posthuman subjectivity entail? In this chapter I will explore the transformations that subjectivity is undergoing in the strands of methodological and radical posthumanism, but also the limitations methodological and radical posthumanism present concerning subjectivity. This discussion will serve as a platform from which to introduce the mediated posthumanist perspective, and delineate a model of mediated posthumanist subjectivity which will move beyond these shortcomings.

Mediated posthumanism borrows largely from methodological and radical posthumanism. With regards to subjectivity, the most important contribution of the latter two lies in the notion that the human being is not an independent and autonomous entity with clear cut boundaries but a heterogeneous subject whose self-definition is continuously shifting, and that exists in a complex network of human and non-human agents and the technologies that mediate between them. At the same time, mediated posthumanism both:

1. Takes up the challenge left unanswered by methodological posthumanism: that which asks what the implications of the breakdown of the humanist separation between subjects and objects are *for human subjects*, not only for technological objects and

2. Offers a critique of the radical posthumanist celebration of hybridity as well as of radical posthumanism's claim that emerging biotechnologies have the potential to bring about a fundamental break with modernity.

In response to these shortcomings, I draw on Foucault's notion of technologies of the self in order to complete a mediated posthumanist model of subjectivity. In such a reading of emerging biotechnologies and posthuman subjectivity, the subject is constituted in specific ways by its technological mediations with the world, but it also develops an active relation to them, so that technologies can be seen as ethical practices that an interconnected, dynamic and molecular subject works with to constitute itself.

6.1   The Liberal Humanist Subject

6.1.1   The Body/Self Connection

It is insofar as subjectivity is tied to *the* body – and to *a* molar body – that a shift in the formulation of bodies can have a direct implication on the formulation of subjectivity. This claim assumes that humans are embodied beings, a notion that has been a central theme of phenomenological inquiry, and extensively treated by Maurice Merleau-Ponty (1962). Embodiment refers to the idea that humans are always located somewhere and at some time and that awareness is profoundly influenced by the fact that we have a body. It implies that the biological and physical presence of our bodies is a precondition for human capacities such as emotion, language, thought, social interaction, and most importantly here, for subjectivity. In his work, Merleau-Ponty raised objections to Descartes' dualism between mind and body. Consciousness, he argues, is not just something that goes on in our heads, rather it is experienced in and through our bodies. The body is not a mechanical object responding to stimuli in its environment, but is constantly in interaction with the world, experiencing, acting and seeking meaning, it is a "phenomenal" body, a "lived" body. Merleau-Ponty illustrates this notion of the lived body with the phenomenon of the phantom limb: if bodies were mere physiological machines, the experience of the phantom limb would not be possible – the machine could proceed without using the limb. But people who experience phantom limbs after having a limb amputated continue to "feel" the limb, and still attempt to use it in situations that call for its use, even though it is no longer there. In the same sense, the lived body is lived in relation to possibilities in the world. For Merleau-Ponty, the body is thus our primary instrument for understanding, and to be a subject means to be in the world as a body. The body should thus be placed at the center of ontology: "I am" because I have a body. It is from the body that I perceive the world. Without a body, I have no place from which to perceive the world. This is also to say, in relation to the present discussion, that a change in body and in one's physical perceptual possibility can transform subjectivity itself.

Foucault, from a very different direction, has examined how changing investments of power in the body have resulted in transformations of conceptions of subjectivity.[1] For Foucault the body is not just the locus of subjectivity, it is the condition of subjectivity. This is because subject production is an effect of power, and power operates and functions through the materiality of the body. In *Discipline and Punish* (1979a), Foucault makes his well-known argument that in the mid-eighteenth century, sovereign power – in which a specific authority defines rule over others – is superseded by disciplinary power – in which techniques of management that cannot be attributed to any particular individual are used to classify and control populations. This new form of power operates on individual bodies, through a constant surveillance that imposes on them a relation of docility-utility. It is this "political technology of the body", power's immediate hold upon the body and its ability to invest it, to mark it, to act on it, to train it and above all to extract knowledge from it, that produces the subject according to Foucault. The modern individual that is produced by this system of disciplinary technologies is at once subject and object, a position in the discourse of knowledge, and also a docile and regulated body.

The body however, is not just the passive, inert target of power's operations in Foucault's work. Even as a site inscribed by technologies of power, it cannot be reduced to these. The body's materiality suggests that it is not entirely tamable and entails a flexibility, an unpredictability, that also endows it with a transgressive potential. Bodies are thus also sites for and instruments of possible resistance to the particular forms power takes, the field on which both power *and* its subversion is played out. For Foucault, attempting to break from the grip of disciplinary powers and the subversion of the construction of normalized subject identities and forms of consciousness requires the reinvention of the body: since the body is so often the target of disciplinary power it should also be instrumental in its resistance.[2] Thus both resistance to disciplinizing power structures and the production of new forms of subjectivity take place via transformations in bodies. In other words, shifts in our dominant formulations of the body will initiate a shift in modes of subjectivity and hence can indicate a shift in historical epistemes.

In both Merleau-Ponty and Foucault's analyses then, while one is much more political than the other, the production of bodies precedes in a certain sense the production of modes of subjectivity. Expanding on Foucault's logic, we can say that if the molar body-as-organism was the locus for modern subjectivity, then the move

---

[1] See especially *Discipline and Punish* (1979a), *A History of Sexuality, Vol. I* (1979b), "Body/ Power" (1980) and "Truth and Power" (1984). Foucault's exploration is a continuation of Nietzsche's focus on the body as the site of the subject's social production and of Nietzsche's genealogy as an analysis of the ways in which history affects or inscribes bodies. See "Nietzsche, Genealogy, History" (1977).

[2] Thus Foucault's rallying cry for new forms of "bodies and pleasures" has inspired a number of projects that take the body as a source for political resistance, from sadomasochist practices to gay bodybuilding. On the other hand, a number of feminist critics have noted that Foucault seems to contradict himself here by first claiming that everything is historically constituted within power relations and then privileging some realm of the body as a transcendental source of transgression. This is namely in relation to his claim in *The History of Sexuality, Vol. 1*, that "bodies and pleasures" are the "rallying point for the counterattack against the deployment of sexuality", implying that bodies and pleasures are somehow "outside" the deployment of sexuality. For this critique, see especially Nancy Fraser (1989: 60), Elizabeth Grosz (1994: 155) and Judith Butler (1997).

to molecular corporeality heralds a shift to a new mode of subjectivity. This would mean that the transformation of classical biological models of bodies, organisms and evolution that we have seen in the previous chapter has far-reaching implications for the autonomous, bounded, unitary notion of self central to the modern episteme. This point is skillfully illustrated by Dorion Sagan, who has co-written a number of books with Lynn Margulis, the leading researcher on endosymbiosis who was referred to at length in the previous chapter. In "Metametazoa: Biology and Multiplicity" (1992), Sagan argues that the molecular, or what he calls "biocentric", view of the body that is emerging with contemporary theories of endosymbiosis and gene-trading bacteria is giving rise to a model of subjectivity that opposes the unitary self assumed in traditional, zoocentric biology.

In the endosymbiotic model of the body as a massive microbial ecosystem, a compilation of chimerical cells, the organism/environment dichotomy breaks down, since, as an organism's connections to its environment grow, that environment becomes part of its body. And this model, according to Sagan, sets the basis for a new kind of subjectivity. Sagan writes, "The body is not one self but a fiction of a self built from a mass of interacting selves. A body's capacities are literally the result of what it incorporates; the self is not only corporal but corporate" (370). The reformulation of the body as an "elaborate mosaic of microbes in various states of symbiosis", is resulting, according to Sagan, in a breakdown of the medically proper animal body and an opening up of the "zoological 'I'" – the encased model of self entailed by the molar body-as-organism – to a radical revision. Sagan concludes,

> The boundaries of selfhood are expanding. In the microbial ecology, the "I" is literally a figure of large numbers. Pieces of the self – from plasmid and viruses to laboratory-spliced genes and prostheses, from milking machines to mechanical and real hearts – are obvious examples of a circulation of elements of subjective identities always already undergoing active (de)composition. (379)

6.1.2 Heidegger Contra the Dualist Paradigm

The idea that a shift in the formulation of the body must be accompanied by a shift in the formulation of subjectivity is crucial to shaping a better understanding of the profound (technologically-driven) transformations of our age. Yet, generally speaking, it seems that the claim for a renewed model of subjectivity meets much greater resistance than the claim for a renewed model of corporeality. This seems to be another expression of that striking inconsistency, discussed in Chap. 2, that characterizes both the narratives of transhumanism and early cybernetic theory: the dissonance between the "openness" of bodies, understood as open to technological enhancement for the former and as feedback systems for the latter, as opposed to the "enclosedness" of the self, which continues to be perceived as a singular entity operating with localized agency. Namely, this is because despite the essential correlation between body and self that I have briefly argued for here, such approaches still uphold a distorted mind/body dualism that allows for transformations in corporeal models that can essentially leave the model of self unchanged.

The critique of the assumption that the human is defined by its separation from the world, that it has an interiority that is set off against the exteriority of an objective, outside world, was developed at length by Martin Heidegger. For Heidegger (1962), the model of the subject as a fixed, self-aware entity that grounds experience, is a misconception – introduced by Descartes and sustained by the philosophical tradition ever since – that fails to get at the fundamental question of what it means to exist. According to Heidegger, this question, the question of the nature of being, is prior to any other structures of human life such as subjectivity, and it is only within such a framework that we can define our place in the world. Heidegger writes:

> In the course of this history certain distinctive domains of Being have come into view and have served as the primary guides for subsequent problematics: the *ego cogito* of Descartes, the subject, the "I", reason, spirit, person. But these all remain uninterrogated as to their Being and its structure, in accordance with the thoroughgoing way in which the question of Being has been neglected. (1962: 44)

Beneath the level of the artificial subjectivities chosen by Western philosophers, Heidegger posited a unique kind of human Being, *Dasein*, that is constituted by its being in the world, by its engagement with worldly objects. Thus the fundamental mode of being is not that of a subject, which defines the human as essentially separated from a world of objects, but of *Dasein*, a complex and open-ended interconnection with the world.

In Heidegger's reading, the human being cannot be conceived as a fixed and independent entity since there is no identity prior to interaction with the world; indeed, it is precisely the interrelations with the world that constitutes human being. The humanist subject thus requires an explicit act of separation that distances humans from their world. In the essay "The Age of the World Picture" (1977b), Heidegger presents this act of separation as the metaphysical ground for the foundation of modern science and hence as the essence of the modern age. The modern scientific method, he argues, transforms truth into the certainty of representation, and objectifies the world, an "objectifying of whatever is … that aims at bringing each particular being before it in such a way that man who calculates can be sure, and that means be certain, of that being" (1977b: 127). In this pivotal development, the world comes to be experienced, conceived and grasped as a *picture*, in terms of an ordering that accords with human categories and needs, and the human being becomes a subject, a being to whom and for whom all that exists must be represented. Indeed, Heidegger argues, the origin of the word subject can be traced back to the Greek *hypokeimenon*, a term that designates "that-which-lies-before, which, as ground, gathers everything onto itself".

This twofold and simultaneous development, by which the world becomes a representation, a world of objects, and the human becomes a knowing subject who experiences the world as a picture, is the radical novelty of the modern age – radical because the modern world picture is not merely the transformation of an older, medieval or ancient, world picture into a modern one, but because the modern age is distinguished by the very fact that the world becomes picture. World pictures and the position humans take in relation to them, in other words, did not exist in medieval or ancient times. This is to say that the subject, as that which assumes that the world stands at its disposition, is not the true essence of human beings, which has finally been freed from the shackles of religious, physical and hierarchical constraints, but a by-product, an illusory effect, of *one of* the possible configurations of the relation between humans and reality: the conceptualization of the world as picture. In the modern era, according to Heidegger, not only has this configuration of the relations between humans and reality become the only valid one, but it renders it extremely difficult to take into account the many ways in which humans and their world are interwoven.

Heidegger's critique of modern metaphysics' rigid separation between subjects and objects acts as a backdrop for the two main approaches, methodological and radical posthumanist, that shape the theoretical background of mediated posthumanism.[3] Here I propose to take a look at these approaches again in terms of subjectivity. As I shall argue, it is in great part due to the limitations in their models of subjectivity that these approaches need to be complemented by a mediated posthumanist perspective.

6.2 The Anthropocentric Perspective Reversed

6.2.1 Symmetry, Mangling and Mediation: Methodological Posthumanism and Subjectivity

In contemporary philosophy of technology and STS, as discussed in Chap. 4, the instrumentalist view of technology, as something that does no more than serve its users' goals, is taken to be misguided. Technological artifacts are seen as affecting human behavior, decision-making and values in ways that challenge this rather simplistic understanding of technology and demand a rethinking of the notion of agency as the sole privilege of human subjects. This idea, that artifacts have some kind of agency, is still perhaps best conveyed in Langdon Winner's rhetorical question "Do Artifacts Have Politics?" (1980). Here Winner describes how the construction of low-pass bridges built over the roads leading to the beaches on Long Island effectively prevented racial minorities and the poor from accessing the beach, since public busses, the main means of transportation used by these groups, could not pass beneath them. It may come as no surprise that artifacts such as those designed by urban planners in big cities do indeed "have a politics", as Winner argued. But the

---

[3] More precisely, many poststructuralist philosophers depart from Heidegger to achieve a critical distance from him, some going as far as claiming that his critique of metaphysics is itself a repetition of an original metaphysical gesture, the gathering of thought to it its "proper" essence and vocation, see namely Derrida's *Of Spirit: Heidegger and the Question* (1989). For theorists like Derrida, something even more original than Being, difference and alterity, is assumed but forgotten by the tradition. Still, Heidegger's analysis of modern metaphysics as world picture is nonetheless a vital – if not radical enough – inspiration for poststructuralism.

idea that artifacts have some form of agency, political or other, extends to relatively "naïve" and simple technologies too, as STS scholars have been persuasively pointing out in an ever-growing list of case studies. Artifacts prescribe users how to act when using them (Akrich 1992). Not only as signs or carriers of meanings that have been written into them by their designers, but as material things that actively transform the "programs of action" of users. In order to take seriously the complex role of technology in society, then, the agency of technological artifacts must in some way be accounted for. The "missing masses" as Latour (1992) has called them, which have been overlooked, socially, politically and philosophically, even as humans interact and depend on them in their everyday lives, need to be acknowledged. This neglect, for many methodological posthumanists, is the result of our humanist ontology, which turns a blind eye to the vast variety of hybrid mixings of humans and non-humans.

Like Heidegger, Bruno Latour views modernity as a process of purifying subjects and objects that begins with the question of the certainty of knowledge about that world for which humans, as conscious subjects, become the sole guarantor. He writes:

> Descartes was asking for absolute certainty from a brain-in-a-vat, a certainty that was not needed when the brain (or the mind) was firmly attached to its body and the body thoroughly involved in its normal ecology. … Only a mind put in the strangest position, looking at a world *from the inside out* and linked to the outside by nothing but the tenuous connection to the *gaze*, will throb in the constant fear of losing reality … (1999: 4, emphasis in original)

Ever since the "strange invention of an outside world" (1999: 3), he contends, the world has been iconoclastically shattered into two ontologically separate categories: sentient, moral and purposive human subjects, and the inanimate objects, characterized by a lack of all the above, that serve them. In this dualist worldview the relationship of the human to the non-human is a purely instrumental one and agency can only be the prerogative of humans. But what the analysis of human-technology relations demonstrates is that any a priori distinction between humans and non-humans cannot be upheld. A more fruitful approach to the study of technoscientific practices should include a dimension of symmetry, by which both humans and artifacts, subjects and objects, are seen as "actants" that have a symmetrical effect on each other. As we have seen, the notion of symmetry works by bracketing off the essentialist nature of entities (as "subjects", "objects", "nature", "culture") in order to focus on how entities engage, connect and associate with each other within networks. There is a shift of emphasis here away from either of the actants in a network – and more specifically away from the subject as that which employs a technological artifact – to a *new* composite entity that is constituted by the engagement between both.

Latour's (1999) example of the NRA's slogan "Guns don't kill people, *people* do", referred to in Chap. 4, offers a simple illustration of symmetry at work. Latour argues that the image, suggested by the NRA slogan, that a shooting involves two distinct entities, a gun understood as a neutral object, and a person, understood as an acting subject, is deceiving. The encounter between subjects and objects, people and guns, argues Latour, evidently has the effect of completely transforming each of these entities, since a gun-holding human is not at all the same as an unarmed human (transformation from "good citizen" to "bad citizen"), and a pointed gun is not at all the same as a gun stored away in an armory (transformation from "silent gun" to "fired gun", or from "sporting gun" to "weapon"). The encounter between these so-called subjects and objects disrupts the functions of subjectivity and objectivity that they were originally allocated, and transforms them into hybrid actors, a new collective which can only act in concert, not in isolation:

> If we study the gun and the citizen as propositions, however, we realize that neither subject nor object (nor their goals) is fixed. When the propositions are articulated, they join into a new proposition. They become "someone, something" else. ... These examples of actor-actant symmetry force us to abandon the subject-object dichotomy, a distinction that prevents the understanding of collectives. It is neither people nor guns that kill. Responsibility for action must be shared among the various actants. (1999: 179–180)

In Latour's symmetry, what seem to be passive non-humans are transformed into actants that modify and are modified by the humans they come into relationship with. This applies to simple materializations such as speed bumps, guns and door-stoppers (1992), and to more complex phenomena like Pasteur's discovery of microbes (1988). The network is the result of an ongoing series of associations and connections between entities of various natures, whose identity is defined only in relation to the other actants in the network.[4]

In the framework of actor-network theory, it is clear that the artifact has come a long way from its categorization as a lifeless, passive, intentional-less object at the service of human subjects. It has programs of actions delegated to it by its designers, moral and normative control, "scripts" and inscriptions. But what of the human? Has it also come a long way from its categorization as an autonomous, individual subject, a "brain-in-a-vat"? As per the principal of general symmetry, the answer has to be yes: such changes in the understanding of the nature of objects necessarily entail a transformation in the understanding of the nature of the subject. If there is no analytical distinction between human and non-humans, if archetypical human properties can in principle be transferred to non-humans, namely "intentionality" and "purposeful action", than what remains of the essential attributes of the subject? For all practical purposes, nothing. Latour writes, "relations of human and non-humans are so intimate, the transactions so many, the mediations so convoluted, that there is no plausible sense in which artifact, corporate body and subject can be distinguished" (1999: 197). Such an understanding of networks leads us away from the possibility of *any* kind of essentialized, isolated or autonomous entity – be this a subject or something else. As with Heidegger, the categorization of one entity or another as subject or object is the outcome of a specifically modern way of ordering reality, not an empirically verifiable occurrence.

Other STS theorists and philosophers of technology also offer frameworks for the analysis of human-technology relations that attempt to overcome the distinctively modern schema of subject-object dissociation. For Andrew Pickering (1995, 2005)

---

[4] Actor-network theory is a type of "material semiotics" – borrowing from semiotics the idea that signs have meaning only in relation to other signs.

humans, non-humans and discursive entities such as theories and conceptual structures interact in such ways that each partner is integrally involved with the other. Pickering is mainly interested in the role of materiality in scientific practice and the process of revision of scientific goals, theories and experimental settings in light of the problems or obstacles that the material world poses to a scientific model. This dialectic, of what he calls "accommodation" and "resistance", constitutes Pickering's notion of the "mangle of practice", a muddle of constantly shifting relationships between humans, machines, theories and instruments that interact in unpredictable ways and through which scientific knowledge is produced.

In such mangles, agency is not restricted to human entities but is also a property of non-human entities, that partake in what Pickering eloquently calls the "dance of agency" that moves in an evolving dialectic (2005). This analogy is not used solely for the production of scientific knowledge, but applies, like in Latour's symmetrical networks, to any association, connection or interaction of human and material beings which respond to one another. For Pickering, like Latour, the recognition of this dance of human and non-human agency shifts the unit of analysis from either things (traditionally the disciplinary realm of the hard sciences) or people (traditionally the disciplinary realm of the soft sciences) to a new kind of posthuman object that dwells at the interface of people and things, at the zone of intersection between these.[5] This new posthuman unit of analysis does not, nor do any of its components, take on the stable, essentialized nature that is inherent to subjects and objects in the humanist paradigm. It evolves and is constantly changing in an unforeseeable fashion. The "posthuman object", Pickering writes,

> has a quality that the traditional sciences lack – it becomes; it does not display the atemporal regularities that physics, ecology or sociology like to look for... This shift exposes a genuine posthuman object which lies, as it were, orthogonally to more traditional objects of enquiry along at least two axes: it is a unity that spans what are usually held apart – the human and the non-human – and this unity is essentially temporal: the coupling of the human and the non-human is situated in time, in the dance of agency, rather than manifesting itself in atemporal laws or regularities. (2005: 35)

Thus, the potency of Pickering's choice of the word "dance" lies not only in its poetic value. A dance always requires that both of its participants be active, but that they loose a portion of (what seems to be) their autonomous agency as they work within a framework of response to one another, cooperation and generation. What's more, it is only within such a dynamic structure that agency is produced. As with Latour, the asymmetry between subjects and objects is not an a priori distinction but only one that is a specific means of describing reality.

---

[5] For example, Pickering (2005) refers to a case reported about Asian eels that were imported to the US as pets for domestic aquariums, that soon began to grow rapidly, climb out of their tanks and invade local waterways. For Pickering, the interesting endeavor here is to think of the people and the Asian eels simultaneously. Not in isolation, but as bound up with one another in an evolving dialectic in which the people imported the eels, then the eels grew and climbed out of their tanks, then the people transferred the eels to the ponds, then the eels began to successfully compete with local fish, etc.

Don Ihde (1990, 1993a, b), by incorporating a phenomenological approach to the human-technology relationship as well as a hermeneutics of technology and culture, also challenges a simple subject-object distinction in his work. Ihde favors the phenomenological method because, rather than offering a description *of* the world, it attempts to describe one's relationship *with* the world, be it in terms of "consciousness" (Husserl), "being-in-the-world" (Heidegger) or "perception" (Merleau-Ponty). In a phenomenological perspective, human beings always experience the world around them, and conversely, the world can only have meaning for human beings in light of this experience, this relationship. Humans and their world are always interrelated in this view; they are in a relationship of mutual self-constitution. Ihde's rendition of the phenomenological method is somewhat unusual in that it rejects more typical readings that see phenomenology as a subjectivist, introspective and intuitive approach, and to avoid confusion between his own approach and this kind of phenomenology, he has proclaimed himself a "post-phenomenologist" (1993b, 2003), where post-phenomenology is,

> (a) … neither subjectivist nor objectivist, but *relational*. Its core ontology is an analysis of interrelations between humans and environment (intentionality). (b) It is not introspective, but *reflexive* in that whatever one "experiences" is derived from, not introspection, but the "what" and "how" of the "external" or environmental context in relation to embodied experience. And (c) all "givens" are merely indices for the genuine work of showing how any particular "given" can become intuited or experienced.[6] (2003: 133)

Within this relationship a phenomenological philosophy of technology, as we have seen in Chap. 4, thus emphasizes the notion of *mediation*, of the mediating role that technological artifacts take on in the relationship between humans and their environment. The notion of mediation emphasizes the ways in which technological artifacts help shape the ways in which humans experience reality.

Peter-Paul Verbeek (2007), continuing the phenomenological work of Ihde, for example, illustrates how images and visualization technologies mediate our experience of the world. Verbeek discusses three models, "modern visions", "postmodern visions" and "posthuman visions", that each concede varying degrees of importance to mediation. In the modern model, which assumes a subject/object separation, images provide an objective relation to reality, and the only mediating role accorded to visualization technologies is as that which determines how objects can be presented to subjects and how subjects can be present in an objective world. Such technologies play a "neutral" role. In the postmodern and posthuman models, the mediating role of visualization technologies is much more obvious in that they present, or translate,

---

[6] In this same essay Ihde explains further:

> Why post? Because, while a pragmatically bonded phenomenology retains the emphasis upon experience, there is neither anything like "a transcendental ego" nor a restriction to "consciousness". Because a pragmatically bonded phenomenology evokes something like an "organism/environment" notion or interactionism, a notion I have repeatedly used as well. Because the *relativity* of pragmatist and phenomenological analyses (not relativism) is a dynamic style of analysis which does not and cannot claim "absolutes", full "universality", and which remains experimental and contingent. (136)

a reality that is not visible to the naked eye. Here technologies are active generators of representations of reality, so that reality is co-shaped by the instrument of perception and the observer. Examples of postmodern visualization technologies include radio telescopes, or the simultaneous use in medical diagnostics of ultra-sound, CT and MRI scanning techniques. The posthuman model puts an even greater emphasis on mediation than the postmodern model insofar as it allows for more intentionality on the part of visualization technologies. Technologies do not just translate aspects of reality that are then pieced together by a postmodern subject; they are more literally the *creators* of reality. Here the technologies involve "artifactual intentionalities" rather than just "human intentionalities stretched over technological artifacts".[7]

For Verbeek, the postmodern and posthuman models, which offer a phenomenological reading of the human-technology-world relationship, assume a radically different understanding of subjectivity. In his examples, reality and subjectivity arise from an interplay between humans and non-humans as it is mediated by technology. Furthermore, in a schema where humans and the world are related via technology, subjects can no longer be seen as entities possessing fixed essences since they are part of constantly changing relations with other humans and non-humans. This is the case in the model of posthuman visions to an even greater extent: while the postmodern model questions the autonomy of the subject because its world is mediated by technologies, it remains a human-centered approach, insofar as the human still "edifies" reality on the basis of the fragments of reality that are presented to it by visualization technologies. In the posthuman model, the intentionality of technologies is no less significant than human intentionality, and the final representation of reality is not pieced together by the observer but produced by the technologies themselves.

The methodological posthumanist perspective thus proposes a move beyond subjectivism and realism. By demonstrating that humans are always implicated in complex socio-technical assemblages, these theorists argue not only for a "stretching" of human intentionality over artifacts, as Verbeek calls it, as that which can be delegated to artifacts by designers and users, but also for an actual extension of intentionality, that becomes a property of artifacts as well as humans. For these theorists, the reality we live in consists of a complex web of relations between the human, the world and the technologies that mediate between them, a network of human and non-human entities that is constantly in the making, constantly creating new realities based on the novel connections and associations being made. In light of such interrelationships, the modern separation of objects and subjects can no longer be upheld. Here subjects and objects emerge as the products, not the prime movers, of the interplay between humans and non-humans.

---

[7] Examples of this are works of art that present aspects of the world that would be impossible to view without specific mediation. Verbeek discusses Wouter Hooijmans' works, where landscape photos are taken using shutter times of several hours which exclude fleeting incidents – animals crossing the field of vision, movements of leaves, etc. – from the final take, creating a reality that is, so to speak, stripped bare of transient occurrences; a reality that does not exist before the camera generates it.

It is clear that the freestanding intentional subject of humanism and the Enlightenment cannot survive this posthumanist rearrangement of principal players unimpaired. It is not so clear, however, what kind of subject this post-subjectivist, posthumanist subject is. That is, if the acknowledgment of an agency of artifacts necessarily implies some degree of deconstruction of the subject/object dichotomy, there is still disagreement on *how much* symmetry is a good thing, that is how much of the subject's "subjectivity" should be relinquished before getting lost in a seamless, monistic web. Pickering (2003), for example, wants to hold on to a form of asymmetry between humans and non-humans, and allows for a stronger type of intentionality among humans. While this asymmetry should not be seen as an a priori distinction, he argues, it is still useful in describing reality. Ihde (2003) also opposes a full-fledged symmetry in which non-humans are actants in the same way that humans are. While subjects and objects are admittedly transformed in the post-phenomenological worldview, they should not be completely eliminated, he argues, to avoid the temptation to either mechanize or socialize the totality – a reductionism that is characteristic of both modernist and symmetrist positions, he adds.

Even in the radical redistribution of agency proposed in ANT, the notion of symmetry pertains mainly to a *functional* equivalence: neither humans nor non-humans have agency as a pre-established essence, to be sure, since agency emerges from within relationships. But it is when humans and non-humans contribute together to constituting a network that the differences between them in terms of agency are erased, that they become functionally equivalent. As Michel Callon and John Law emphasize: "Yes, there are differences between conversations, texts, techniques and bodies. Of course. But why should we start out by assuming that some of these have no active role to play in social dynamics?" (1997: 168). Latour does not argue that humans and non-humans are the same, but that they have equal capacity to enter into novel combinations and collectives, in which agency is shared. As McMaster and Wastell (2005: 17) point out, "There is no crude argument in Latour that humans and machines are the same, no talk to be found anywhere of Turing tests, no contention that machines per se have human intelligence, spiritual aspirations or are actuated by moral impulses". The point of emphasis for Latour is less a reconceptualization of human subjectivity then the argument that humans and non-humans have equal capacity to enter into novel combinations and collectives, and that this means that we must aim to create "well-articulated" collectives, where all members and their mutual entanglements can be made visible for debate. One could argue that the significant factor is not so much that humans and non-humans be treated symmetrically, but that they are defined relationally in the network.

These internal disagreements set aside, there seems to be a real lacuna in discussions that involve the meaning of post-subjectivist and post-realist subjectivity among these theorists.[8] Does the agency of technologies imply that they are active in the same way that humans are? If the notion of symmetry is actually quite confined, applying solely to the function of entities in a network, and further limited to

---

[8] Verbeek's (2011) attempt to develop an account of a mediated, ethical subject is exceptional in this sense, and I will return to his adoption of Foucault in this context later on.

specific networks at specific moments in time, than does it really have any conceptual weight? Does an extension of agency and intentionality to artifacts completely collapse the subject/object dichotomy, or just disturb it a little? Does this have ramifications for essentialism per se, or just for thinking about technological artifacts? These questions need to be asked more persistently by methodological posthumanists. The shift to non-dualist and non-essentialist frameworks of analysis that the methodological posthumanist perspective calls for has inevitable implications for human ontology that methodological posthumanists often seem reluctant to explore.

It seems that the most plausible reason for this is that the attempt to take materiality seriously, to reinstate materiality as a central feature of human and social activity, involves for these theorists a focus on the immediate and the immanent, rather than on the enduring and the transcendent. For contemporary philosophers of technology as we have seen, this entails a move away from the broad abstract theorizations of technology "with a big T" that dominate classical philosophy of technology, and towards a new emphasis on nuance, pragmatism and empirical analysis of specific technologies. For STS scholars, this entails a move away from deterministic understandings of technological development and an emphasis on the diversity of explanations – cultural, social, institutional – of technological development, that is brought to the fore in case studies that describe *how* more than *why* sociotechnical collectives act. In both disciplines, this can be seen as a shunning away, or a "stepping down", from metaphysics and ontological aspects in general, that reflects a "wariness of the large-scale claims common in social theory" (Law 2009: 142). For these theorists, especially proponents of ANT, it is the presupposition of essences within human-technology networks that is problematic. There is no good reason to assume, from the outset, that humans play a more important or the only role in social dynamics (even if we might reach this conclusion post-analysis).

In this sense, the unwillingness of methodological posthumanists to pursue the implications of non-human agency for humans can be seen as a result of self-imposed epistemological constraints. But it can also be seen as an unwillingness to delve into a nascent metaphysics, and itself as a kind of performance of the non-essentialism that they uphold. Thus, it may not be fair to require of methodological posthumanism that it conceptualize a coherent understanding of posthumanist subjectivity (nor a full-blown relational ontology for that matter). But it is nonetheless regretful that it seems to stop short of this, and perhaps even intentionally; that, having in a sense breathed life into objects, we are left guessing what happens to the subjects who have had to relinquish their humanist privileges.

6.2.2 The Subject as Effect: Poststructuralist and Radical Posthumanist Subjectivity

Radical posthumanism, as discussed in Chap. 2, is greatly influenced by the antihumanism of poststructuralist theory and postmodern theory, especially in terms of its views on subjectivity, so that it is interesting to briefly recall what happened to the subject under poststructuralist and postmodern scrutiny.[9] Poststructuralist and postmodern theory's critical position vis-à-vis the notion of subjectivity is not obvious from the outset. In its early days, it was rather an *excess* of subjectivism, in the sense that objectivity collapsed into personal subjective preferences, that seemed to be the hallmark of new poststructuralist ideas. But as with Heidegger and methodological posthumanism, it is the critique of objectivity, here in the form of a skepticism towards claims to a singular, objective truth that were revealed as myths, narratives or social constructions, that inevitably led to a questioning of the subject as it was constructed by the modern tradition. For postmodern theory, however, this necessary turn to the subject takes on a much greater emphasis than it does in methodological posthumanism.

As the grandest narrative of all, the autonomous subject as origin, telos or center of intentional action, became a main target for poststructuralist critique, which set out to reveal that the subject is not an a priori category but is produced through language and systems of meaning and power. That subjectivity is a construct, an effect, became a core assumption of poststructuralist thought, while theorists vary in what they see to be the *processes* by which individuals are constituted as subjects and given unified subject positions. For Jacques Lacan (1977), for example, the identity of the ego is illusory, and the forces responsible for the construction of identity are always beyond the grasp of those constituted by them.[10] Subjectivity emerges with the individual's entrance into the "symbolic" of language. Louis Althusser (1992), in an attempt to create a Marxist theory from which all traces of human agency could be expunged, argues in his famous 1970 essay "Ideological State Apparatuses", that the idea of a unified human agent is an illusion fostered by ideology, so that the subject is an *effect* of ideology, not the other way around. In terms of society, this means that the subject does not exist before society but is a contingent effect of it. For Jacques Derrida (1976), the deconstruction of objectivity must be accompanied by a deconstruction of subjectivity. Like Lacan's adoption of structural linguistics, Derrida sees subjectivity as dependent on, or arising out of, language. His insistence on the primordiality of difference necessarily undermines the idea of a unitary subject. Difference, or *différance*, does not just imply that there are differences between subjects – this would still assume that there is a subject "who" differs, whose existence is prior to difference. Rather it is difference that precedes and allows for subjectivity.

As we have seen in the beginning of this chapter, Foucault (1979a, b, 1989) conceives the subject in relation to specific discursive practices; thus a large part of his

---

[9] Structuralist theory already had unsettling implications for the notion of the autonomous subject. Claude Lévi-Strauss (1963), in his adoption of de Saussure's linguistic model to anthropology, argued that if all of culture is structured like language, than meaning is reducible to a system of differences, and along with meaning, agency and history are reduced to consequences of structure.

[10] Lacan's work can be seen as a reaction to the tendency towards an "ego psychology" developed in post-Freudian psychoanalysis. In an amusing reworking of the Cogito he writes: "I think where I am not, therefore I am where I do not think … I am not wherever I am the plaything of my thought; I think of what I am where I do not think to think". (1977: 166)

work is dedicated to identifying various ways in which claims of truth intersect with structures of power to articulate forms of human subjectivity. The subject here is an effect, not a source, of knowledge. In his genealogical works, Foucault shows that subjectivity has not been the same for every epoch, and that forms of subjectivity are determined by the rationality embedded in the discursive practices of the times and the subject-positions they articulate. The modern individual subject, more specifically, is produced by the accumulation of a corpus of knowledge about human beings and a complex of overlapping disciplinary technologies. Modern institutions, from the prison through the school to the hospital, are all involved in the disciplining of bodies through techniques of surveillance that transfix the individual and produce a subject that reflects the norms of power.

Gilles Deleuze and Félix Guattari (1977, 1987) are particularly interested in how the psychoanalytical account of the consciousness produces modern subjects who conform to law and authority through the containment and channeling of libidinal energy into molar stratifications. The illusion of the individual subject, they argue, is fostered by systems of repression that channels the overlapping and often contradictory investments of desire into unifying or totalizing "molar" investments. Subjects are multiplicities that are formed as a result of many different investments of desire in the social field; and it is the subject's "desiring machines", those arrangements of heterogeneous parts by which the flow of energy is produced and cut, as we have seen, that occupies the central position, not the ego. Thus the subject in Deleuze and Guatarri's works is not only decentered, it is no longer recognizably anthropocentric, since "desiring-machines" – much like the networks and manglings of methodological posthumanism – are formed by the coupling of both human and non-human parts, with no distinction drawn between them. Non-human machines, such as social, cultural, environmental and technological assemblages, are as much as inter-personal relations and intra-familial complexes, the source of subjectivity. Theirs is perhaps the most radical rejection of the modern subject, against which they propose a "schizo" or "nomad" subject who seeks to resist the capitalist axiomatic, reject Oedipus and break through representational identity into the realm of becoming.[11]

The postmodern decentering, deconstruction or dispersal of the subject implies two inversions; two reversals of meaning that subvert the humanist model of subjectivity. First, not only is subjectivity constructed rather than constitutive, an effect of modern discourses and institutions rather than their origin (and here the claim made by methodological posthumanism that human/non-human couplings precede the subject/object split parallels the postmodern claim, minus the postmodern emphasis on discourse, semiotics and language). But secondly, the process of subject formation in postmodern analysis is not the result of progressive liberations as in the Enlightenment narrative, a dialectical unfolding that traces the evolution of human liberty. Rather, it is a process of repression, imposition and coercion. Althusser's

---

[11] Of course, numerous other postmodern theorists and their analyses of modern and postmodern subjectivity could be cited here. Jean Baudrillard (1983b) claims that subjects have imploded into the masses and that in the postmodern media and information society subjects are no more than "a term in a terminal" (1983a); KroKer and Cook (1988) view the subject as a cyberneticized effect of systems of control; and Frederic Jameson (1984) argues that the postmodern subject is no longer bounded, centered or has any psychic depth.

(1992) popular illustration of ideological subject formation, for example, presents a policeman who, by hailing a passerby, "hey you!", interpellates a subject, "creating" a "you" as the passerby accepts this status by responding to his call. The terms of subjective recognition are an imposition from the outside. For Foucault, the subject is also interpellated, by the church, the school and the state. In Foucault's account, the subject is formed by power's imposition on individuals followed by the internalization of the discourse of power which comes to constitute the subject's self-identity. Subjectivity, then, incorporates two contradictory meanings, as both a condition of agency and a subordination, in the sense of being "subject to". The subject is both subjected to forms of individuation shaped by the demands of power, and to practices that fix each individual with a known, stable identity. The elucidation of subject formation as a repressive rather than a liberatory process lies at the basis of the postmodern political valorization of the dissolution of the subject. In addition, and no less important for the political implications of postmodern theory, is the recognition that modern subjectification always implies the objectification of others, who are excluded from the narrow and particular caste of individuals that "universal" subjectivity designates.[12]

The emphasis in postmodern theory is less on formulating an overall theory of subjectivity than on deconstructing the modernist notion of a unified, rational, autonomous subject who is the result of a process of liberation. It is here that the familiar attack on postmodern theory as a philosophy that rejects and destructs all things modern without offering any productive alternatives easily finds it place. The absence of a coherent theory of subjectivity – or what it is that replaces subjectivity – however, is not just a methodological shortcoming, but a logical consequence of the critique of modern subjectivity, since it is the very fixity and determinedness of identity that renders modern subject formation repressive. Hence it is the practice of dissolving the modern subject itself that should make possible the emergence of new types of decentered subjects that can resist the process of subjection modern subjectivity implies. In this sense, the deconstruction of modern subjectivity is an emancipatory practice, that can liberate some form of authentic subjectivity from the terror of fixed and unified identity, allowing it to become free, dispersed and multiple. Postmodern subjectivity emerges not as *a* postmodern subject, but as the right to difference, variation and metamorphosis.

6.2.3 Radical Posthuman Subjectivity: Hospitality and Prostheticity

The radical posthumanist model of subjectivity incorporates both the postmodern critique of modern subjectivity and much of the philosophy of technology and STS critique of the subject/object split. What makes it particularly interesting is its

---

[12]Feminist and postcolonial theory, more than any other offshoots of poststructuralism, have argued that what masquerades as universal subjectivity has been reserved for white European males.

combination of the latter's move beyond subjectivism and humanism toward materiality, i.e. the idea that subjectivity is also an effect of technological mediation, and not just of social and discursive structures, *and* the former's political valorization of the dissemination of the subject and its emphasis on the ethical nature of the construction of new kinds of subjectivities. But if we take Heidegger's thought as a kind of starting point for this type of thinking about technology and subjectivity, as I have suggested, then we can also see how far radical posthumanism has taken the critique of modern subjectivity, or how uncompromising its model for posthuman subjectivity is.

As we saw have seen, in "The Question Concerning Technology" Heidegger overturns the conventional subject/object relationship on its head by presenting the notion of "Man" as voluntarist subject and master of an objectified nature as an effect rather than the origin of the instrumentalist view of technology. The portrayal of the human being as subject and the world as object is a consequence of technology establishing itself, Heidegger argues, not the other way around. This means that his critique of modern technology does not transpire from the belief that "Man" as subject objectifies everything, exploiting and dominating his surroundings for his own satisfaction, but from the belief that, taken to its logical conclusion, the model of modern technology will result in the *very disappearance* of subjects and objects once they are both transformed into standing-reserve. When the goal of technology becomes the increasingly efficient ordering of resources, simply for the sake of ordering, objects, followed by subjects, begin to dissolve:

> when man, investigating, observing, pursues nature as an area of his own conceiving, he has already been claimed by a way of revealing that challenges him to approach nature as an object of research, until even the object disappears into the objectlessness of standing-reserve. (1977a: 300)

It is precisely this dissolution, for radical posthumanism, that should be celebrated. For radical posthumanists, the posthuman subject is a radically new mode of subjectivity, characterized by heterogeneity, openness and variation, "a cluster of complex and intensive ... assemblages which connect and interrelate in a variety of ways" (Braidotti 2006a: 16). No longer unitary, self-evident and coherent, it is "unstable, multiple and diffuse" (Poster 1990), an "amalgam, a collection of heterogeneous components, a material-informational entity whose boundaries undergo continuous construction and reconstruction" (Hayles 1999: 3); a "multiple and fragmented entit[y]" (Stone 1995). It is "hybrid" (Haraway 1991), "nodular" (Taylor 2001) and "transversal" (Parisi 2004); fluid and irreducible to a single dynamic, "polyphonic and heterogenetic" (Guattari 1995). Instead of being limited by its bounded organism-barrier, it is open to its surroundings, indeed, it is its relationality with what would be considered the bounded organism's "outside" or "other" that constitutes this ex-centric, non-anthropocentric posthuman subject.

I referred to the ethical implications of the outward orientation of the body in Deleuze's Spinozist account of the body in the previous chapter. In light of the claim for originary prostheticity, I argued, the extroversion of the molecular body is seen as something essentially positive. This ethical implication takes on a much greater significance when this extroversion of the body entails that the subject, too, is unbounded and extroverted. Thus, radical posthuman subjectivity can also be situated within the tradition of the "ethics of alterity"; a detail which suggests the importance of the meeting point between molecular corporeality, originary prostheticity and the positive engagement with technology assumed by radical posthumanism.

For Emmanel Levinas (1969), who has developed one of the most significant contributions to approaching alterity, ethics always occurs in relation to the other, where the other is that which cannot be reduced to the domain of the same. In Levinas' thought, the self starts off as a self of enjoyment, a happy, satisfied being that "enjoys" sensations and journeys into the world to make everything other part of itself. The ethical moment arises when this self encounters an other – something that it wants to enjoy but that it cannot because the other resists consumption. It is thus the incalculable alterity of the other that creates the conditions for the emergence of an ethical moment since, because the other cannot be made part of the self, it demands to be addressed, or reacted to. For Levinas, this "proximity" of the other, that comes from the outside, is responsibility (1981: 139), or the ability to respond, and it allows for the possibility of ethics. But this moment is not only the grounds upon which ethics shows itself, it also gives rise to a new subjectivity, or rather, it is from this confrontation that subjectivity arises. For Levinas, subjectivity is born from this encounter, one that indicates that the self is a projection towards the other. Insofar as subjectivity arises from the confrontation with the other, it can thus be seen as intrinsically prosthetic, and as entirely different from the humanist discourse of subjectivity.

In the context of technology, the molecular body and the posthuman subject's openness to technology, that is, to non-human otherness, can be seen as a reworking of Levinas' ethics of respect for alterity, and can be interpreted as an ethical act, insofar as it acknowledges and welcomes alterity. The philosopher Joanna Zylinska (2002), for example, uses Levinas' notion of an ethics of respect for the other to develop what she calls a "prosthetic ethics of welcome". For Zylinska, the Levinasian idea of the encounter with the other can be articulated in the idea of the prosthetic relationship, where prosthesis is portrayed as "an ethical figure of hospitality, of welcoming an absolute and incalculable alterity that challenges and threatens the concept of the bounded self" (217). Zylinska applies this framework of philosophy of alterity in the realm of technology and "technological others", and argues that it offers a productive model for thinking about the human as relational and co-emerging with technology, as always already enhanced in the sense of originary prostheticity.

This is to say that the ethical opening of radical posthuman subjectivity does not only involve an opening *outwards*, it is also a welcoming or invitation *inwards*, an ethical openness to what Derrida (2000) has called "unconditional" or "absolute" hospitality.[13] Just as in Levinas' account the approach of the other is a source of astonishment, of "catching off guard", Derrida's unconditional hospitality assumes that one is unprepared for the unexpected arrival of an other, and that one ask nothing of the other, even if this deprives one of the mastery of one's space and one's home. Of course, Derrida is more concerned with actual human, flesh and bones guests in a context of nation-states and immigration. But in the context of technology, often posited as the radical other of the human, and in terms of selfhood, hospitality implies a danger of the intrusion of the unknowable into the confined territory of the self, and demands that rather than reinforcing those boundaries, one welcome an alterity that threatens the very idea of self-mastery.

For Zylinska, the conceptualization of ethics in terms of hospitality provides a model of open-ended ethical responsibility that can account for an affinity of humans not only with traditional human others and animals, but also with technology. Furthermore, this terminology of hospitality ties in quite interestingly with the discussion in the previous chapter on endosymbiosis, the process by which one organism is incorporated by a *host* organism and fuse over time. In Margulis' explanation of the origin of eukaryotic cells, the mergings of "host" and "parasite" prokaryotic cells can lead to the development of novel metabolic capabilities and new entities. Similarly, the molecular posthuman subject can be thought of in terms of a host organism that is essentially open to its technological environment. Indeed, Stelarc, the performance artist whose works were mentioned in Chap. 4, has adopted the terminology of hospitality in relation to his performances. "The body", he writes, "has been augmented, invaded and now becomes a host – not only for technology, but also for remote agents" (Stelarc 1997: 66). Thus Zylinska (2002b) discusses his "Stomach Sculpture" project, in which a metal dome sculpture was inserted in his body and followed with the help of endoscopic equipment, as an example of unconditional hospitality. As he opens his body up to the intrusion of technology, she argues, he abandons the desire to "master the house of his own body" (232). "In this context", Zylinska writes, "prosthesis can be interpreted as an ethical figure of hospitality, of welcoming an absolute and incalculable alterity that

---

[13] Extending Levinas' ethics of obligation towards the alterity of the other, in *Of Hospitality* (2000) Derrida posits hospitality as the name for our relation to the other, the very principle of ethics. For Derrida, the notion of hospitality includes two types of hospitality, "general" and "unconditional" hospitality. General hospitality makes claims to property ownership and the desire to a form of self-identity. It implies that, in order to be hospitable, one must first have the power to host, i.e., one must be the "master" of one's house (country or nation); general hospitality assumes a necessary degree of control over the situation. Unconditional, or absolute hospitality, on the other hand, demands a welcoming of whomever, or whatever, may be in need of hospitality, hence involving a relinquishing of control in regard to who will receive that hospitality. This hospitality necessitates a "non-mastery" and the abandoning of all claims to property or ownership. According to Derrida, the unconditional form of hospitality is near impossible to enact, but, nonetheless, the idea of hospitality is inconceivable without it. The co-existence of these two different and conflicting notions of hospitality gives rise to an aporia, an internal tension that is precisely, he claims, what keeps the concept of hospitality alive.

challenges and threatens the concept of the bounded self" (217). The human does not disappear here so much as it emerges as dependent and co-evolving with technology in a way that calls the sovereign humanist subject into question and depicts it in an unethical light.

6.2.4 Radical Posthuman Subjectivity: Strategic Resistance

Radical posthuman subjectivity, by creating new possibilities of connection between the self and its others, thus implies an essentially ethical component since it calls into question the relationship between the self and its others and can potentially lead to a fundamental repositioning of human beings in relation to both technological and organic environments. This ethical overtone is reinforced by the significance of radical posthuman subjectivity as a *strategic* position, a platform from which to resist power. Radical posthuman subjectivity, with its assumption of originary prostheticity, calls into question the "ontological purity" (Graham 2002) of modern, molar subjectivity. In political terms, this is cause for celebration for radical posthumanists insofar as it deconstructs oppressive boundaries that have defined what the normative human should be in modernity, thus posing a powerful threat to patriarchal capitalism (Haraway 1991). Braidotti writes:

> This post-humanistic acceptance of hybridization and the intermingling of the biological with the cultural, the physical with the technological is neither nihilistic nor decadent. Nor is it a romantic valorization of otherness per se. It is rather an attempt to disengage the process of becoming from the classical topos of the dichotomy self-other and the notion of "difference" from its hegemonic and negative implications. (2006b: 170)

In this sense, radical posthumanism incorporates both the attack on the dialectical form of modern thought and the binary logic of Self and Other that underlie modern discourses of patriarchy, racism and colonialism, as well as the insistence on difference, fragmented identities and hybridity as a means of contesting such systems of domination. Radical posthumanism can thus be seen as a continuation, or expansion of an entire host of modern and contemporary liberation struggles into the realm of science and technology.

More specifically regarding subjectivity, if the understanding of identity as a fixed category is seen as a method of political control, of producing manageable, tractable subjects, then the hybrid nature of posthuman subjectivity is inherently subversive. In her well-known essay "Split Subjects, not Atoms, or How I Fell in Love with my Prosthesis", (1995) Sandy Stone argues that at the current dawn of the virtual age, terms such as distance, direction and presence are becoming increasingly problematic as subjects are becoming multiple, fragmented entities. "A disembodied subjectivity messes with whereness", she explains. "In cyberspace you are everywhere and somewhere and nowhere, but almost never *here* in the positivist sense" (1995: 398). As the decoupling of the agency/body schema proceeds in virtual systems, she argues, agency is not only delegated, but dissolved, and governments must increase and intensify the means for tracking citizens. Such "location" or "warranting" technologies, by which governments have traditionally maintained order (credentials like social security numbers, passports, addresses, the Diagnostic and Statistical Manual), attempt to prevent the gradual "disappearance" of socially and legally constituted individuals. In light of the increasing pressure applied by location technologies, Stone proposes a celebration of cyberspace as a non-Cartesian mode of location, where it becomes increasingly difficult to track dissolving, fragmenting subjects who refuse to be one thing by choosing to be many. For Stone, prosthesis is the extension of the self that can defy warranting.

The notion that cyberspace is a kind of non-place of resistance where contemporary power formations can be challenged is a common theme among radical posthumanists. Katherine Hayles defines the "revolutionary potential" of virtual reality as that which can "expose the presuppositions underlying the social formations of late capitalism and open new fields of play where the dynamics have not yet rigidified and new kinds of moves are possible" (1993: 175). Mark Poster views cyberspace as a space where "copyright law, fixed identities, censorship and so forth are continuously evaded and challenged", as a "highly differentiated field of resistance, conflict, and uncertainty" (2004: 328); and Arthur Kroker and Michael Weinstein (1994), who calls Nietzsche the "patron saint of the hyper-texted body", views the wireless body online as a "rebel" against the "virtual class", as well as an expression of Nietzschean affirmation. In these narratives cyberspace is depicted as a non-place where multiple and polymorphous identities can be achieved, where the notion of self takes on a fluid and negotiable nature. Cyberspace is taken to be a privileged site for the performance of posthuman subjectivity, the experimental grounds for the creation of various types of identity, that cybernauts can easily move through (Bukatman 1993; Plant 1995: Turkle 1995).

Richard Doyle (2003), for example, discusses mind uploading – that future technology so cherished by transhumanists – as the mark of a new "ecology" or "topology" of subjectivity that allows for richer concepts of identity. Where his admiration differs significantly from transhumanists, and represents the underlying difference between liberal and radical posthumanism, is that uploading allows precisely for a creation of many selves, rather than a replication of the same self, where this is possible precisely because the self itself in the radical posthumanist approach is always and already a dynamic, changing and emergent property, rather than a fixed essence. Doyle stresses that while programs may be quite predictable, they retain an ontological contingency insofar as they depend on temporal and material factors for their instantiation (138). Uploading will never be, then, an inscription of the same, but always an unpredictable and imperfect replication of different versions of identity at different moments. Doyle cites Robin Hanson in *If Uploads Come First*:

> Uploads who copy themselves at many different times would produce a zoo of identities of varying degrees of similarity to each other. Richer concepts of identity would be needed to deal with this zoo, and social custom and law would face many new questions, ranging from

"Which copies do I send my Christmas card to?" to "Which copies should be punished for the crimes of any one of them?" (in Doyle 2003: 138)

The uploader, or the program that an uploader becomes, is no longer tied to a single, central agency, a univocal self. It does not reproduce an original, but each copy evokes something new.[14]

6.3   Towards a Mediated Posthuman Subjectivity

6.3.1   A Critique of Radical Posthuman Subjectivity

For these theorists, such virtual interactions emphasize the extent to which subjectivity has never been a constant, and can allow for its remaking. But such a celebration of cyberspace, uploading and cloning, and the multiple, fragmented identity they foster is at risk of valorizing disembodiment, by which cyberspace becomes a space unconstrained by the meaning and matter of the corporeal; and a disappearing body is easily associated with the ancient dream of transcending the body and a reinstated Cartesianism.[15] Scott Bukatman (1993), whom I mentioned above, while identifying the empowering potential of cyberspace, sets out precisely to narrate the loss of corporeality in the discourses of much current science fiction and cultural theory, where bodies are so often rendered as occasions for coding. Unwittingly, such a move carries the danger of falling back into the very critique that radical posthumanism holds against transhumanism, with its one-sided emphasis on the

---

[14] Doyle associates such iteration of differing identities with William S. Burroughs' musings on cloning and the ego in "Immortality" (1993). For Burroughs, cloning offers the possibility of multiplying and distributing identity, of deterritorializing the self and scattering it among various bodies. Rather than giving significance to the ego, he views it as the end of the ego:

> What we think of as our ego is defensive reaction, just as the symptoms of an illness-fever, swelling, sweating – are the body's reaction to an invading organism. Our beloved ego, arising from the rotten weeds of lust and fear and anger, has no more continuity than a fever sweat. There is no ego; only a shifting process... When I first heard about cloning I thought, what a fruitful concept: why, one could be in a hundred different places at once and experience everything the other clones did. I am amazed at the outcry against this good thing not only from men of the cloth but also from scientists, the very scientists whose patient research has brought cloning within our grasp. The very thought of a clone disturbs these gentlemen. Like cattle on the verge of stampede [sic], they paw at the ground mooing apprehensively. "Selfness is an essential fact of life. The thought of human selfness is terrifying". ... Terrifying to whom? Speak for yourself. (132)

[15] The critique that digital technologies betray a desire to transcend the body in favor of pure, disembodied information is by now common. See Braidotti (1994, 2002), Hayles (1999) and Ihde (2002), and for a more general engagement with the significance of embodiment, Butler (1993), Grosz (1994), Lingis (1994), and Lakoff and Johnson (1999).

immaterial and the disembodied.[16] Indeed, there is an unmistakable inclination among a number of critical theorists towards a euphoric celebration of virtual embodiments, such as KroKer and Cook (1994) and Jean Baudrillard (1991a, b), that reduce the body to a mere surface of representation.

Feminist theorists, emphasizing women's culturally invested and loaded relationship to both the ideological female and lived material body, are usually quick to pick up on such shortcomings and identify them as reenactments of the Cartesian mind-body dualism.[17] For example, writing against the dangers of a hyped-up disembodiment and speaking as someone who has a *real* prosthetic leg, Vivian Sobchack, evokes the actual pain involved in becoming a cyborg:

> What many surgeries and my prosthetic experience have really taught me is that, if we are to survive into the next century, we must counter the millennial discourses that would decontextualize our flesh into insensate sign or digitize it into cyberspace … Prosthetically enabled I am, nonetheless, not a cyborg. Unlike Baudrillard, I have not forgotten finitude and the naked capacities of my flesh, nor, more importantly, do I desire to escape them. (Sobchack 1995: 209)

Critiques of Stelarc's works have also been advanced along these lines. Stelarc's performances are most often interpreted as postmodern, cyborgian and posthuman explorations of the limits of corporeality and subjectivity that problematize our definitions of the human. But in so doing, Stelarc attempts to demonstrate that the body and the human nervous system are unadapted to the information age and will be superseded in the future of human evolution. In this technofantastic vision, Stelarc is known for claiming that "the body is obsolete", conceived of as an object, an "it", not an "I" (he refers always to "*the* body", rather than "*my* body"). This is a body, explains Mark Dery, that "must be hollowed, hardened, and dehydrated, its essential innards are scooped out so that it can be a 'better host for technology'" (1996: 162). Stelarc's performances and announcements thus seem to be much less radical than they are presented as being, because they do not succeed in escaping the Cartesian mind-body dualism that he claims to contest. In the redistribution of agency that his works seem to perform, there is actually a heightened experience of separation between an observing mind and a docile, manipulable body, and this distorted Cartesianism enhances the fantasy of transcendence of the body through technology.

This is a real problem arising from the celebratory position of radical posthumanists towards new technologies that seems to indicate a deeper problematic that underlies the notion of originary prostheticity and hybridity. Is it really possible to distinguish between supplemental and originary prostheticity? Can hybridity subsist

---

[16]Though, it is precisely *against* such a reading of uploading that Doyle argues. "Uploading", he contends, "is an anticipation of precisely 'more life', life not free of the body but distributed into spaces not yet visible … Addicted to contingency, or prowling for mastery, uploaded subjects mark out the materiality and possibility of the new flesh, anticipated flesh that is something other than either transcendental or meatless" (2003: 142).

[17] Some even claim that the many assumptions of subject reformulation in postmodern thought realign it with preexisting arguments that valorize masculine transcendence of matter as a feminine and inferior principle. See Somer Brodribb's *Nothing Matters: A Feminist Critique of Postmodernism* (1992).

without assuming that the two entities that constitute it were at first separate? The difficulty even of defining the concept of hybridity without recourse to some prior distinction attests to this. As Vicki Kirby (1997) points out, in a skillful critique of Haraway's cyborg figure, in order to fabricate the hybrid and intermingled cyborg, one must first begin with the discrete component entities which are precisely those elaborated within the logic of identity, self and other, human and technology. In other words, in the construction of the cyborg, technologies – even when conceptualized as originary prosthetics – seem to always remain to some extent an appendage, insofar as they must at some point intersect with a "non-technological" human body.[18] This is to say that the logic of hybridity in general indicates the existence of some originary moment of human purity before a "fall" into technics. Kirby writes:

> It is against the unity of "the before", the purity of identity prior to its corruption, that the cyborg's unique and complex hybridity is defined … Haraway's "disassembled and reassembled" recipe for cyborg graftings is utterly dependent upon the calculus of one plus one, the logic wherein pre-existent identities are *then* conjoined and melded. The cyborg's chimerical complications are therefore never so promiscuous that its parts cannot be separated, even if only retrospectively. (1997: 147, emphasis in original)

As long as technological artifacts and processes are applied to bodies and selves – and Haraway does posit these as subsequent in some essential sense to bodies and selves (she writes: "communications technologies and biotechnologies are the crucial tools *recrafting* our bodies", (1991: 164, emphasis added)) – even if a countless number of novel variations can potentially emerge from this encounter, the two categories that proceeded the cyborg conjunction, human and technology, remain largely intact. The cyborg's real irony emerges here as its reinscription into the binary logic of identity and its expression of a Cartesian recuperation.

This is not the only problem that runs through the radical posthumanist model of subjectivity. There is something highly ambiguous about the type of claim that is being made for radical posthumanist subjectivity. On the one hand, the notion of a heterogenous, unbounded, extroverted subject that is mediated by its technologies and is in constant interaction with its environment is an ontological claim about human subjectivity: this claim implies that we have always been posthuman, or what amounts to virtually the same thing, that the posthuman model of subjectivity best describes what it means to be human, in a non-historically specific way. On the other hand, in light of the poststructuralist theories of subject formation that radical posthumanism incorporates, there is no such thing as a subject that is "outside" of history, since each historical epoch is distinguished by a specific model of subjectivity. In this sense, the posthuman subject is a historical contingency, a new mode of subjectivity that emerges from connections between humans and technologies that are qualitatively new.[19]

---

[18] Kirby's critique is in line with Deleuze's attempt to rethink difference not as difference *from* but difference in itself, as well as the difficulty of this endeavor.

[19] In Chapter 2, this problem manifested itself in radical posthumanism's positioning on both the historical and the ontological poles of the second axis of differentiation.

Radical posthumanist discourse tends to alternate between both these claims quite easily, paying little head to the fact that, on the face of it, these two positions are incoherent. On a theoretical level, as we shall see shortly in the mediated post-humanist approach to subjectivity, this incoherence is not highly problematic. But when subjectivity has concrete political implications and is the site for political resistance as it is for radical posthumanism, this inconsistency is a serious short-coming. I will take a short detour to explain this.

In the "Cyborg Manifesto", Haraway interprets the new biotechnologies and proliferating communication systems as key markers of a transition from older hier-archical social structures to a new form of power that she calls the "informatics of domination" (1991: 161). This emerging world order, she claims, transcends the sets of dualisms (nature/culture, public/private) that underpinned the established system of meaning upon which "White Capitalist Patriarchy" has relied for centuries. Contemporary power, she notes, does not work by normalized heterogeneity any more, but rather by "networking, communications redesigns, stress management". Haraway's understanding of the "informatics of domination" is analogous to Gilles Deleuze's (1995) notion of "societies of control". Deleuze suggests that contempo-rary societies are no longer disciplinary in the Foucauldian sense: where subjectivi-ties in disciplinary society were produced within the walls of social institutions such as the school, the factory, the hospital, etc., at present the production of subjectivity is not limited to any specific place, but extends across the entire social terrain. Today, the walls of disciplinary society, so to speak, have come down, and control is continuous and integral to all activities and practices of existence. Likewise, Haraway claims that Foucault's analysis of the disciplining of bodies appears already out of date, indeed, that Foucault named a form of power at its moment of implosion, and that the formulation of biopower needs to be "enterprised up" (1991: 245, fn 4; and in Gane 2006).

Such an understanding of our current society as post-disciplinary is also a main feature of Michael Hardt and Antonio Negri's *Empire* (2000), which provides an extensive analysis of the postmodernized global economy as the current formation of power. While Hardt and Negri cannot be called radical posthumanists per se, they can be associated with radical posthumanism insofar as they discuss the emergence of a posthuman figure, what they call the "new barbarian",[20] as a subject which can resist imperial sovereignty. For Hardt and Negri, imperial sovereignty is also a form of power which is much closer to Deleuze's control society than Foucault's biopolitics: "The production of subjectivity in imperial society", they claim, "tends not to be limited to any specific places. One is always still in the family, always still in school, always still in prison, and so forth" (2000: 197). This is because the current ideology of corporate capital and the world market, they argue, is an anti-foundationalist and anti-essentialist discourse, of which notions like creativity, mobility, diversity and mixture are the very conditions of possibility. Fixed boundaries and binary divisions in this global landscape are impediments to the free circulation and multiplication

---

[20] Hardt and Negri's main protagonist is the "multitude", but I am here interested in their discussion of the posthuman figure who uses technoscience to resist Empire.

of differences, which capital seeks to include within its realm. In this new imperial configuration of power, argue Hardt and Negri, the affirmation of hybridity and the proliferation of difference can no longer be seen as liberatory, since Empire strives to do away with the same modern forms of sovereignty and binary divisions that postmodern theory set out to challenge. Despite their "best intentions", Hardt and Negri powerfully write, postmodernist and postcolonialist theory,

> may end up in a dead end because they fail to recognize adequately the contemporary object of critique, that is, they mistake today's real enemy ... the postmodernist politics of difference not only is ineffective against but can even coincide with and support the functions and practices of imperial rule. The danger is that postmodernist theories focus their attention so resolutely on the old forms of power they are running from, with their heads turned backwards, that they tumble unwittingly into the welcoming arms of the new power. (2000: 137, 142)

Both Haraway and Hardt and Negri thus succeed in identifying and describing a novel structure of post-disciplinary power in their work, one that thrives on hybridity and difference. And yet, in both cases their figures of posthuman resistance – the cyborg and the "new barbarian" – are celebrated for resisting forms of power they have identified as a thing of the past. The political potential of Haraway's cyborg lies in its transgression of the system of binary oppositions that underlies Western patriarchal power apparatuses; its imagery, she claims, can "suggest a way out of the maze of dualisms in which we have explained our bodies and our tools to ourselves" (1991: 181). But the cyborg is precisely the product of an age that is characterized by the *breakdown* of these same oppositions and dualisms. Thus its strategic potential is defined more in relation to its ability to disrupt boundaries that have in any case already given way in the "informatics of domination", than in relation to the contemporary form of power that she identifies.

Similarly, Hardt and Negri argue that resisting imperial civilization requires the creation of new posthuman bodies that would be based on the recognition that the seemingly rigid boundaries between nature, humans and machines are not fixed, and that nature is "an artificial terrain open to ever new mutations, mixtures, and hybridizations" (2000: 215). Hardt and Negri's new barbarian is a subject that embodies the will to be against Empire, and can use experiences of corporeal transformations and permutations as means of resistance. "The contemporary form of exodus and the new barbarian life", they write, "demand that tools become poietic prostheses, liberating us from the conditions of modern humanity" (217). Here also, the new barbarian is hailed as a figure that resists the "conditions of *modern* humanity", who escapes the normalization effects of disciplinary power (216, emphasis added). This posthuman subject will resist by, so to speak, taking the "bio" out of biopower: "The will to be against", they exclaim, "needs a body that is incapable of adapting to family life, to factory discipline, to the regulations of a traditional sex life, and so forth". But the imperial configuration of power in which this body acts – as Hardt and Negri uphold throughout the book – *is no longer* a modern one, nor one that is based primarily on disciplinary subject production.

Another way of understanding this ambivalence is that, in both these examples, the mobile, fragmented, posthuman subject, be it the cyborg or the new barbarian, is simultaneously a *symptom* of postmodern conditions and an *agent of resistance* to postmodern power. Hardt and Negri, for example, first assume a stance against the liberatory potentials of the postmodern figure of multiple subjectivity as we have seen above. But once they establish that such fragmented multiple identity is conservative in the context of empire (and what's more, that empire thrives precisely on these principles), they then uphold this kind of postmodern identity as the site of resistance to empire. This gives rise to a very confusing picture, and this is, I think, an extension of a characteristic problematic of postmodern theory in general, that of the difficulty of distinguishing between postmodern theory's descriptive and prescriptive function.

As opposed to those theorists *of* the postmodern who bemoan the postmodern rupture as a loss of traditional values and certainties (Bell 1976; Toynbee 1963), or as an abandonment of the unfulfilled potential and emancipatory values of modernity (Habermas 1981), postmodern theorists view what they see as a break with modern ideologies and practices as an opportunity to affirm new models of subjectivity, thought, language and society. It is here that postmodern theorists most fundamentally differ from those precursors of postmodern critique or "anti-modern" currents within modernity such as Nietzsche, Heidegger, Bataille or the Frankfurt School. Postmodern theory specifically speaks of an *end* of modernity and demonstrates a sense of new historical possibilities, a new capacity to think outside the framework of modern binaries and modern identities. But this development of new models of thought does not fulfill a simply descriptive role, describing new phenomena as they appear in a new historical era. The new categories of thought formulated by postmodern theory are prescriptive. There is an underlying – albeit often obscured – assumption within postmodern theory that the categories of thought developed offer a *better* account of reality than the categories offered by modern thought: better in the sense of more accurate or correct, and better in the sense that if these models were to replace the discourses and practices of modernity, we would also be "better off" – in the same way that the model of radical posthumanist subjectivity is often assumed by radical posthumanists to offer a better account of what subjectivity is in general, not only in our specific age. This argument cannot but sound vague because of the problematic nature of using normative notions such as "better", "more just", etc., within a postmodern theoretical framework. Nonetheless, postmodern theory does much more than just refer to a historical period, one that follows modernity; it is an approach to understanding reality (better), and hopefully changing it (for the better). It is both an ontological claim about the "real" nature of being and a political project aimed at social transformation according to that understanding.

Granted, Haraway and Hardt and Negri are writing within a tradition that views power as that which creates the conditions for its own resistance, that contains the seeds for its own destruction, which cannot come from without that system of power. For Hardt and Negri, Empire cannot be resisted by returning to any previous social organization, on the contrary, the process of globalization must be accelerated, by pushing globalization beyond its present limitations: "globalization", they write, "insofar as it operates a real deterritorialization of the previous structures of exploitation and control, is really a condition of the liberation of the multitude"

(52). Furthermore, Hardt and Negri do distinguish themselves from Haraway and other radical posthumanists in a significant way.[21] They persistently emphasize the idea that resistance without creation (of a "new life") is in itself not enough:

> In addition to being radically unprepared for normalization, however, the new body must also be able to create a new life. We must go much further to define that new place of the non-place, well beyond the simple experiences of mixture and hybridization, and the experiments that are conducted around them. (216)

This additional requirement seems to be missing among radical posthumanists, and Hardt and Negri should be credited for emphasizing and warning that so long as the will to be against Empire does not lead to the construction of a new body and a new life, it remains on a non-productive level of mere refusal, and risks reinforcing imperial power rather than resisting it. This is because the "methods" of the anthropological exodus they speak of – namely hybridization and mutation – are precisely the methods used by imperial sovereignty itself. In the same way, the liberatory potential radical posthumanists see in new technologies is liberatory in relation to modern forms of power, the liberal humanist subject and the system of binary oppositions which it is based on, not in relation to the new form of power, characterized by the collapse of the nature/culture and nature/technology, that runs through contemporary technoscience. It remains unclear why, acknowledging the emergence of a *new* world order of "informatics of domination" or "imperial sovereignty", these theorists would not conceptualize *new* strategies of resistance.[22]

6.3.2 *Technological Mediation and the Aesthetics of Existence*

The core of my critique of the radical posthumanist model of subjectivity is rooted in the ambiguity that arises from the simultaneous historical and ontological claim that is made for radical posthumanist subjectivity. This bilateral claim becomes inconsistent I argued, when subjectivity is seen as the site for political resistance, i.e., when it takes on a prescriptive more than a descriptive function. But when what is at stake in developing a model of posthuman subjectivity is not a means of political resistance to modernity, the bilateral nature of this claim is much less problematic.

---

[21] They also insist that the actualization of terrains of potential metamorphosis requires new productive/labor practices, rather than cyborg "fables".

[22] Luciana Parisi and Tiziana Terranova (2000), in their analysis of the shift from "discipline" to "control" and its investment in the body, put this even more bluntly:

> Even as discipline was being successfully exported through outsourcing from the West, the relief for its decline was palpable in the early stages of postmodern theory. The latter got drunk on its glimpses of a different age, one based on proliferation, fragmentation, and fluidity and forgot that discipline was a historical formation not the ultimate form of power. Postmodern theory was weak in its understandings of modes of power which did not operate by enclosure, individuation and hierarchy and sometimes misunderstood the collapse of discipline for the end of power as such. (15)

In other words, when a posthumanist approach is not about describing (or prescribing) a rupture with modernity but about offering analytical tools to comprehend the technologically mediated transformations of the present as developments that incorporate *both* "modern" and "postmodern" trends, as I will argue in the next two chapters, both the historical and the ontological claim for posthuman subjectivity can be reconciled, and offer a basis for a mediated posthumanist model of subjectivity. This reappraisal of a non-humanist posthuman subjectivity can act as the basis for a mediated posthumanist model of subjectivity.

In this context, the historical and the ontological claims for a non-humanist posthuman subjectivity are actually two ways of stating the same thing: that subjectivity is always in the making, both historically, in the sense that different cultural and historical periods will be characterized by different models of subjectivity, and ontologically, in the sense that subjectivity, or more generally "being human", is never a static, fixed category but is always an effect of our relationships with others – subjects and objects – in the world. In other words, the quantity of human and non-human connections that we are witnessing today may be a historical novelty, but this reflects an aspect of being human that, although it is perhaps more noticeable today, is not a novelty. Another way of understanding this is that developments in technology make new discursive situations possible, open up new subject positions, and make possible new forms of human being, but only because the subject is *already* intrinsically open to new positions. It can be helpful to understand this kind of posthumanist subjectivity in line with Foucault's later work on ethics and subject constitution, and further of emerging biotechnologies as "technologies of the self", a direction that has recently been opened up by several theorists in the philosophy of technology as well as the sociology of biomedicine (Dorrestijn 2012; Rose 2007; Verbeek 2011).

Foucault's overall project can be understood as a tracing of the history of practices of subject-constitution, where subjectivity is always a product of the interplay between structures of power. Viewed in terms of this overall project, a distinction can be drawn between Foucault's early works, which can be seen as investigations of the forces and structures of power that determine or produce specific subjects via disciplining technologies, and his later works, where he also explored the ways in which subjects can shape relationships with structures of power, by which they can contribute to their own subject-constitution. In this later period, represented mainly by the second and third volumes of *The History of Sexuality* (1985, 1986) and a number of seminars and interviews (1993, 1997a, b), Foucault employed historical research to show how subjectivity is not only a coercive, but also a *formative* power of the self, that can be experimented with.

This distinction or shift of emphasis is outlined in his seminar entitled "Technologies of the Self" (1997b). Here Foucault set out a typology of four interrelated "technologies", or ways that human beings develop knowledge about themselves and that each imply a certain mode of training individuals: technologies of production, technologies of sign systems, technologies of power or domination and technologies of the self. Foucault's work largely focuses on these last two, technologies of power and technologies of the self, the former involving the exercise of force on human bodies while the latter pertain to the formation of thinking, feeling and acting subjects. Together, these technologies are said to comprise the phenomenon of "governmentality", those practices whose aim is the governance of the lives of individuals. Having hitherto focused on technologies of power and domination, those technologies that objectivize subjects, Foucault proposes, at this later point in his work, to study the technologies of the self, or technologies of individual domination, in their own right.[23]

As part of this shift of emphasis, Foucault develops in these later works a new mode of ethical thought, that he describes as an *aesthetics of existence*. Foucault distinguishes two primary elements of the ethical domain: moral codes – overarching rules that can regulate and prescribe how to act in various situations, and relations to the self – or the forms of subjectivation, the various ways that subjects "subject" themselves in accordance with a moral code. While all moralities contain both elements, they are seen as mutually exclusive. Thus, where there is an emphasis on codification, on systematicity and on the richness and capacity of codes to extend to all areas of behavior, the relations of the self to itself are depreciated. And vice versa, in a morality in which forms of subjectivation and practices of the self are stronger, the exact observance of moral codes becomes less important (Foucault 1985: 30). Foucault identifies the malaise of modern ethics as a symptom of an over-emphasis on codification, which essentially decreases the individual's margin of freedom, and he looks to the ethics of classical Antiquity as an ethics which was centered on the self, in which the relation of the self to itself was vital. The history of ethics from late Antiquity to modernity in this sense can be seen as a substitution of practices of the self with moral codes.[24]

The ethical approach that Foucault discovered in classical Antiquity is explicitly directed at constituting oneself as a specific subject. Its main focus is the question of what kind of subject one wants to be, not the question of how to act or which moral rules one should follow. For example, in Christian ethics, Foucault argues, sexual activity was forbidden in most forms (and severely restricted in the rest) according to a strict moral code that viewed sexual acts on the whole as evil. The Greeks, conversely, viewed sexual acts as natural and necessary goods, although subject to abuse, and emphasized the proper use or government (*chresis*) of pleasures, with proper moderation. Sexuality here was organized primarily in terms of styling. The rules of self-control and discipline applied to sex allowed access to pleasure and to truth, and was a means for shaping the self into the kind of ethical subject that could manage a household and participate in the government of the city. Similarly, medicine for the Romans in the first two centuries AD was not conceived as just a set of interventions and remedies in the case of illness. It was seen as a way

---

[23] To the extent that technologies of the self can be separated from technologies of power. For Foucault all four technologies are at work simultaneously.

[24] In "Foucault as Virtue Ethicist", Neil Levy (2004) claims that insofar as virtue ethics' main thesis is that modern ethics has placed too much emphasis on rules, duties and consequences, when the core of ethics should be the character of the moral agent, it can be seen as a parallel project to Foucault's.

for the self to work on itself, "a corpus of knowledge and rules, a way of living, a reflective mode of relation to oneself, to one's body, to food, to wakefulness and sleep, to the various activities, and to the environment" (1986: 100). Medicine offered technical means by which the self could relate to itself, to family members and more broadly to society, and could experience itself as an ethical subject.

Ethics is thus a matter of "care of the self": "in order to behave properly, in order to practice freedom", writes Foucault, "it was necessary to care for self, both in order to know one's self … and to improve oneself" (1987: 5). The issue of precedence here, from care, to knowledge, to improvement of oneself, is significant. For the Greeks, Foucault explains, care of the self was a precondition of that other important Delphic precept, to "know yourself". Self-knowledge, the access to truth, could only take place concurrently and in view of a caring of self:

> in and of itself an act of knowledge could never give access to the truth unless it was prepared, accompanied, doubled, and completed by a certain transformation of the subject; not of the individual, but of the subject himself in his being as subject. (Foucault 2005: 15)

Care of the self was also a precondition for being able to care for others. First, in the Socratic-Platonic form, as an activity that is required in order to know how to govern well (in this sense "caring" for others refers to governing others properly). And later in Epicurean and Stoic philosophy as a permanent obligation for every individual throughout life. Foucault illustrates this process or dynamic within ethics as a kind of circle, that goes from the self as an object of care, to the knowledge of government as the government of others (2005: 39).

The passage from an ethics centered on practices of the self to a code-based morality that Foucault diagnoses can then be understood as a certain reversal of this process, or at least a shift of emphasis away from care of the self, that happened in two "moments". First, in Christian ethics concern for the self came to be interpreted as egoistic self-love, as something immoral, that should be replaced with the concern that one must show for others, self-renunciation and selflessness (taken up in modern ethics as the obligation towards others, as other people, collectives and classes). Secondly, with the Cartesian moment's elevation of knowledge – grounded in the self-evidence of one's own existence – as the fundamental means of access to truth. The dissociation of care of the self and knowledge of the self, and the forgetting, so to speak, of the importance of the former in order to achieve the latter that transpires from this Cartesian moment, marks the beginning of the modern age of the history of truth for Foucault, and the detachment of the "philosophical theme" from the "question of spirituality" (2005: 17).[25]

It is insofar as ethics is concerned with a care of the self for the Greeks, that the practices of self, or technologies of the self, are considered ethical practices, be they discursive (aimed at self-knowledge) or concerned with the body, such as diet, health, sexual activity or household management. Technologies of the self are defined as the various technologies,

---

[25] It also marks, at least since the Enlightenment, an inextricable link between the demand to know yourself, to access your inner truth, and the constitution of subjects who can be governed.

which permit individuals to effect by their own means or with the help of others a certain number of operations on their own bodies and souls, thoughts, conduct, and way of being, so as to transform themselves in order to attain a certain state of happiness, purity, wisdom, perfection, or immortality. (1997b: 225)

Foucault's use of technologies of the self refers primarily to existential techniques, not technologies in the sense that we refer to biotechnologies, digital technologies, or even technological artifacts. But it is a common aspect of those emerging biotechnologies that have prompted debates about posthumanity and the nature of human nature, that their use and development is taking place on a backdrop of new understandings of how health, well-being and the good life are being construed: namely in terms of the achievement of these goals via an active engagement on the part of individuals with technologies. As will be discussed in more detail in the next chapters, these technologies engender new guidelines, norms and ideals according to which subjects think about and understand themselves, and according to which they act. As such they can also be called technologies of the self.

A framing of emerging biotechnologies as technologies of the self, as practices of the self on the self, suggests that they are transformative and transfiguring in very powerful ways. Ways that neither the instrumental nor the substantive models of technology that were discussed in Chap. 4 can account for. For the Greeks, according to Foucault, access to the truth, the possibility of "knowing yourself", necessitated a prior or at least corollary progressive transformation of the self via the practices of the self. In order to become capable of "truth", subjects had to subject themselves to a certain preparation of "work of the self on the self, an elaboration of the self by the self, a progressive transformation of the self by the self for which one takes responsibility in a long labor of ascesis" (2005: 15). There could be no truth without conversion or transformation of the subject first. This is the question of what the subject must be in order for the subject to have access to truth. But, further, the knowledge of self that could then be accessed, in turn, is transformative of the subject – the subject is improved in this process and can then turn care towards others. This is to say that aspects of the subject are also transformed by virtue of this access to the truth, that this is a reflexive phenomenon. Foucault calls it the "rebound effect": "The point of enlightenment and fulfillment, the moment of the subject's transfiguration by the 'rebound effect' on himself of the truth he knows, and which passes through, permeates and transfigures his being" (2005: 18). In these terms, the problem with codified moralities, and with the dissociation of care from knowledge of the self in the Cartesian moment, is precisely that this "rebound effect", the transformative essence of ethics, and of knowledge, can no longer exist. Access to truth loses its transfigurative impact and aims at no more than the indefinite development of knowledge.

The rebound effect of technologies of the self is precisely what is implied in the notion that the subject is mediated by technologies, by the technologies of the self that emerging biotechnologies are today. This can also be understood in other terms. Subjectivity in Foucault's work is constituted as an effect of power relations – it is mediated. Power relations in Foucault's work, as I recalled earlier in this chapter, need not be only thought of in terms of ideological frameworks.

Material artifacts, such as the classic example of the Panopticon, play a central role as subject-constituting forces. We can also say that technology, particularly in a highly technological culture, can be seen as one of the sources of power in relation to which subject-constitution takes place – so that subjectivity is also technologically mediated. But it is Foucault's attention to the subject's *relation to this relationship*, as the primary target of his ethics, that is particularly useful for thinking through the hopes and fears raised in posthuman discourse. The term aesthetics of existence implies that ethics is a matter of stylizing those relationships to the powers, drives and impulses that govern the self, of giving them shape. It is crucial that in Foucault's ethical work one can develop a certain distance from the relations of power that constitute the subject, a distance that allows one to fashion a productive relationship to those relations.[26] This opens up an essential space of freedom in Foucault's work. Not in the sense of the liberation of some true inner nature or essence *from* power. There can be no such thing, insofar as what may be taken to be a "true essence" or "inner nature", an authentic subjectivity, is the *result* of power relations, not something that stands in separation to and against them. Rather, freedom here is the possibility of modifying the impact of power on one's subjectivity, it is a practice of actively engaging with one's relationship to power and so a practice of subject constitution. Freedom is not about escaping structures of power but of interacting with them. Because there is no authentic or natural self that can be liberated, freedom lies in the dynamic, aesthetic and experimental self-creation undertaken in the practices of the self; technologies of the self are practices of freedom.

This mode of freedom situates the subject somewhere between being completely independent and autonomous in relation to the world around it and being completely determined by the structures of power that make up that world. This is a very constructive starting point from which to think about how humans can relate to the technological mediations that help constitute them. It can allow for an approach that remains critical to technology and its influence on our lives, without technology becoming a deterministic force from which we need to be protected, or, as the flip side of this, an approach that can allow for the identification of specific relationships to technologies that are enriching without this requiring that we embrace all technologies indiscriminately. As Verbeek (2008b, 2011) has argued, this kind of consolidation of Foucauldian ethics and technological mediation can be the grounds for a very fruitful framework in the ethics of technology, based in a more positive or productive articulation of the relationship to power. Here, the subject actively engages with the technological mediations that help constitute it. Its actions are not simply the result of a technological determination but of an active appropriation, or the "stylizing" of these mediations. This kind of subjectivity proceeds along an aesthetic program that continuously enriches its relation to the world.

---

[26] This echoes Nietzsche's work on morality, in which human beings are also, in a natural state, governed by impulses and passions (as manifestations of the will to power). The essence of morality for Nietzsche is the aim of disciplining, mastering these drives, of transcending the merely given. Morality is not about the repudiation or abnegation of these impulses, but about sublimating them, about giving "order to the chaos" by stylizing one's character and self-overcoming.

This is a pragmatic ethics, in the sense that it articulates a way of "dealing" with technology beyond the "take it or leave it" dispositions of instrumental and substantive approaches. It is also essentially non-humanist. Rather than starting off with a strict separation of the human and the technological and then positing ethics as that which protects the human from the technological, it indicates that from within the many points of interweaving of the human and the technological – moments of technologically mediated subject constitution – mediation can be guided in desirable ways. For Verbeek (2011) this implies that the central question of this kind of ethics of technology will be what kind of mediated subjects we want to be, what kind of subjectivity is implicitly organized by the mediating role of specific technologies. In the next two chapters I will try to extend or complement this view by identifying novel conceptualizations of nature and subjectivity that engagements with these technological mediations are engendering. In the examples of the notions of "nature" in the context of assisted reproduction, "life" in the genomic sciences, "authentic selfhood" and "subjectivity" in the context of neuroscience and the new genetics, we shall see that emerging biotechnologies are contributing to the idea that these categories, like subjectivity in Foucault's aesthetics of existence, are both given and given to control, determinative, or transformative, but not deterministic. It is the difficulty to articulate this ambiguity, and richness, which is most distinctive of emerging biotechnologies insofar as they echo something about what it means to be human. And it is the space that is opened up, of freedom, in our engagements with them that should be the focus of our ethical and philosophical explorations.

6.4 Conclusion

In this chapter I tried to provide a thorough account of what posthuman subjectivity entails for methodological posthumanism, radical posthumanism and mediated posthumanism, suggesting that the latter is a more viable model of the posthuman subject in light of several shortcomings in the former approaches. This mediated posthumanist model of subjectivity begins, like methodological and radical posthumanism, with the assumption that the autonomous, fixed and unitary subject is a by-product of the rigid separation of subjects and objects undertaken by modern metaphysics, rather than some true essence of human beings. This understanding that the division of the world into subjects and objects is only one specific configuration of the relations between humans and the world, and quite an inadequate one at that, allows for other configurations that may take into account the many ways in which humans and their world – their so-called objects, their environment and their technology/ies – are interwoven.

Methodological posthumanism shares this assumption, but is interested more in its implications for the agency of objects than in its implications for a new ontology of subjects. On the other hand, methodological posthumanism makes an important contribution to posthumanist discourse by bringing to the fore the significance of materiality and technological mediation as an aspect of human being that must be accounted for if we attempt to understand the importance of technology in our present age. Radical posthumanism also shares this critique of the subject of modern metaphysics and contributes to posthumanist discourse a combination of methodological posthumanism's emphasis on materiality with two aspects of poststructuralist theory: the ethical significance that is implied in the construction of new kinds of subjectivities and the political valorization that emerges from the dissemination of the autonomous, unitary subject. But as we have seen, the significance for radical posthumanism of subjectivity as a strategic position, a platform from which to resist power, gives rise to an incoherence by which it is not clear how the multiple and fragmented nature of posthuman subjectivity, which can understandably act as a site of resistance to modern disciplinary power, can also embody the ideal form of resistance in a post-disciplinary configuration of power that is itself multiple and fragmented. In this context it is necessary to question what qualitative kind of impact hybridity, multiplicity and fluidity really have.

To help move beyond these limits, I suggested that Foucault's later focus on the subject's relation to the relations of power through which subject-constitution takes place, his approach to ethics as the stylizing of the self via self practices, can be very useful. Namely, insofar as it contributes to a consolidation of non-humanist subjectivity and the notion of technological mediation, and opens up the possibility of relating to technologies, i.e. of "guiding" subject-constitution, in desirable ways. In this framework, ethics does not center on the autonomous moral agent who stands in opposition to a technological world from which it must be protected or which it must learn to manipulate in order to enhance that autonomy, but on the practices that constitute human beings as moral subjects. Emerging biotechnologies can be construed in this framework as technologies of the self, as practices that are deployed by individuals upon themselves in order to transform themselves in desired ways.

Finally, this chapter on subjectivity necessarily opens up onto a larger inquiry. If along with new models of corporeality, new models of subjectivity are contesting the model of the autonomous, free-standing and bounded subject of modernity, does this indicate a broader shift? That is, if as Foucault argued in *The Order of Things*, the humanist subject "man" was the product of the modern historical episteme, than can we maintain that we are currently entering a new "postmodern" or "posthuman" episteme? According to radical posthumanism we can. As Haraway writes,

> If belief in the stable separation of subjects and objects in the experimental way of life was one of the defining stigmata of modernity, the implosion of subjects and objects in the entities populating the world at the end of the Second Millennium … are stigmata of another historical configuration. (1997: 42)

Indeed, as I have suggested, the belief that recent technoscientific developments have the potential to usher in the end of modernity is one of the defining characteristics of radical posthumanism. But it is precisely on this issue that mediated posthumanism and radical posthumanism diverge most significantly. As I will argue in the next chapter, while we are currently witnessing new models of nature in the context of technoscience that *do* indicate a move beyond humanism, too many elements characteristic of modernity persevere at present that challenge Haraway's forecast. Rather, modern and "postmodern" tendencies coexist simultaneously, intertwine and give rise to novel notions of nature and the human, that challenge the very dichotomy between modern and postmodern.

References

Akrich, M. (1992). The de-scription of technical objects. In W. Bijker & J. Law (Eds.), *Shaping technology/building society* (pp. 205–224). Cambridge, MA: MIT Press.

Althusser, L. (1992). Ideology and ideological state apparatuses. In A. Easthope & K. McGowan (Eds.), *A critical and cultural theory reader*. Toronto: Toronto University Press.

Baudrillard, J. (1983a). The ecstasy of communication. In H. Foster (Ed.), *The anti-aesthetic: Essays on postmodern culture*. Port Townsend: Bay Press.

Baudrillard, J. (1983b). *In the shadow of the silent majorities*. New York: Semiotext(e).

Baudrillard, J. (1991a). Ballard's "Crash". *Science Fiction Studies, 18*(55). Available at http://www.depauw.edu/sfs/backissues/55/baudrillard55art.htm. Accessed 22 August 2013.

Baudrillard, J. (1991b). Simulations and science fiction. *Science Fiction Studies, 18*(55). Available at http://www.depauw.edu/sfs/backissues/55/baudrillard55art.htm. Accessed 22 August 2013.

Bell, D. (1976). *The cultural contradictions of capitalism*. New York: Basic Books.

Braidotti, R. (1994). *Nomadic subjects: Embodiment and sexual difference in contemporary feminist theory*. New York: Columbia University Press.

Braidotti, R. (2002). *Metamorphoses: Towards a materialist theory of becoming*. Cambridge: Polity Press.

Braidotti, R. (2006a). Affirming the affirmative: On nomadic affectivity. *Rhizomes, 11/12*(2005/2006). http://www.rhizomes.net/issue11/braidotti.html. Accessed 13 Jun 2013.

Braidotti, R. (2006b). *Transpositions: On nomadic ethics*. Cambridge: Polity Press.

Brodribb, S. (1992). *Nothing matters: A feminist critique of postmodernism*. North Melbourne: Spinifex.

Bukatman, S. (1993). *Terminal identity: The virtual subject in postmodern science fiction*. Durham: Duke University Press.

Burroughs, W. S. (1993). Immortality. In *The Adding Machine: Collected Essays*. New York: Arcade Publishing

Butler, J. (1993). *Bodies that matter: On the discursive limits of "sex"*. New York: Routledge.

Butler, J. (1997). *The psychic life of power: Theories in subjection*. Stanford: Stanford University Press.

Callon, M., & Law, J. (1997). After the individual in society: Lessons on collectivity from science, technology and society. *Canadian Journal of Sociology, 22*(2), 165–182.

Deleuze, G. (1995). *Postscript on control societies* (trans: Joughin, M.). *Negotiations: 1972–1990*. New York: Columbia University Press.

Deleuze, G., & Guattari F. (1977). *Anti-Oedipus: Capitalism and schizophrenia* (trans: Seem, M., Lane, H.R., & Hurley, R). New York: Viking Press. Original edition, 1972.

Deleuze, G., & Guattari, F. (1987). *A thousand plateaus: Capitalism and schizophrenia* (trans Massumi, B.). Minneapolis: University of Minnesota Press. Original edition, 1980.

Derrida, J. (1976). *Of grammatology* (trans: Spivak, C.G.). Baltimore: Johns Hopkins University Press. Original edition, 1967.

Derrida, J. (1989). *Of spirit: Heidegger and the question* (trans: Bennington, G., & Bowlby, R.). Chicago: University of Chicago Press.

Derrida, J. (2000). *Of hospitality* (trans: Bowbly, R.). Stanford: Stanford University Press.

Dery, M. (1996). *Escape velocity: Cyberculture at the end of the century*. New York: Grove.

Dorrestijn, S. (2012). Technical mediation and subjectivation: Tracing and extending Foucault's philosophy of technology. *Philosophy and Technology, 25*(2), 221–241.

Doyle, R. (2003). *Wetwares: Experiments in postvital living*. Minneapolis/London: University of Minnesota Press.

Foucault, M. (1987). The ethic for care of the self as a practice of freedom. In J. Bernauer & D. Rasmussen (Eds.), *The final Foucault*. Cambridge: MIT Press.

Foucault, M. (1977). Nietzsche, genealogy, history (trans: Bouchard, D.F.). In D. F. Bouchard (Ed.), *Language, counter-memory, practice*. Ithaca: Cornell University Press.

Foucault, M. (1979a). *Discipline and punish: The birth of the prison* (trans: Sheridan, A.). New York: Vintage Books.

Foucault, M. (1979b). *The history of sexuality, volume 1: An introduction* (trans: Hurley, R.). London: Allen Lane.

Foucault, M. (1980). Body/power. In C. Gordon (Ed.), *Power and knowledge: Selected interviews and other writings, 1972–1977* (pp. 55–62). New York: Pantheon Books.

Foucault, M. (1984). Truth and power. In: P. Rabinow (Ed.), *The Foucault reader* (pp. 51–100). New York, Pantheon.

Foucault, M. (1985). *The use of pleasure* (trans: Hurley, R.). New York: Pantheon.

Foucault, M. (1986). *The care of the self* (trans: Hurley, R.). New York: Pantheon.

Foucault, M. (1989). *The order of things: An archaeology of the human sciences*. London/New York: Pantheon.

Foucault, M. (1993). About the beginning of the hermeneutics of the self: Two lectures at Dartmouth. *Political Theory, 21*(2), 198–227.

Foucault, M. (1997a). The ethics of the concern for self as a practice of freedom (trans: Hurley, R.). In P. Rabinow (Ed.), *Ethics, subjectivity and truth: The essential works of Michel Foucault 1954–1984, vol. 1* (pp. 281–301). New York: The New Press.

Foucault, M. (1997b). Technologies of the self. In P. Rabinow (Ed.), *Ethics, subjectivity and truth: of the essential works of Michel Foucault 1954–1984, vol. 1* (pp. 223–251). New York: The New Press.

Foucault, M. (2005). *The Hermeneutics of the subject: Lectures at the Collège de France, 1981–1982* (trans: Burchell, G.). New York: Palgrave Macmillan.

Fraser, N. (1989). *Unruly practices*. Minneapolis: University of Minnesota Press.

Gane, N. (2006). When we have never been human, what is to be done? Interview with Donna Haraway. *Theory, Culture and Society, 23*(7–8), 135–158.

Graham, E. L. (2002). *Representations of the post/human: Monsters, aliens and others in popular culture*. New Brunswick: Rutgers University Press.

Grosz, E. (1994). *Volatile bodies: Towards a corporeal feminism*. Bloomington: Indiana University Press.

Guattari, F. (1995). *Chaosmosis: An ethico-aesthetic paradigm* (trans: Bains, P., & Pefanis, J.). Bloomington: Indiana University Press.

Habermas, J. (1981). Modernity versus postmodernity. *New German Critique, 22*, 3–14.

Haraway, D. (1991). A cyborg manifesto: Science, technology, and socialist-feminism in the late twentieth century. In D. Haraway (Ed.), *Simians, cyborgs and women: The reinvention of nature* (pp. 149–181). New York: Routledge.

Haraway, D. (1997). *Modest_Witness@Second_Millenium. FemaleMan©_Meets_Oncomouse™: Feminism and Technoscience*. New York: Routledge.

Hardt, M., & Negri, A. (2000). *Empire*. Cambridge, MA: Harvard University Press.

Hayles, N. K. (1993). The seductions of cyberspace. In V. Conley (Ed.), *Rethinking technologies*. Minneapolis: University of Minnesota Press.

Hayles, N. K. (1999). *How we became posthuman: Virtual bodies in cybernetics, literature and informatics*. Chicago: University of Chicago Press.

Heidegger, M. (1962). *Being and time* (trans: Macquarrie, J., & Robinson, E.). New York: Harper and Row. Original edition, 1927.

Heidegger, M. (1977a). The question concerning technology. In D. F. Krell (Ed.), *Martin Heidegger: Basic writings* (pp. 287–317). New York: Harper & Row.

Heidegger, M. (1977b). The age of the world picture. In W. Lovitt (Ed.), *The question concerning technology and other essays* (pp. 115–154). New York: Garland.

Ihde, D. (1990). *Technology and the lifeworld: From garden to earth.* Bloomington: Indiana University Press.

Ihde, D. (1993a). *Philosophy of technology: An introduction.* New York: Paragon.

Ihde, D. (1993b). *Postphenomenology: Essays in the postmodern context.* Evanston: Norhtwestern University Press.

Ihde, D. (2002). *Bodies in technology.* Minneapolis: University of Minnesota Press.

Ihde, D. (2003). If phenomenology is an albatross, is postphenomenology possible? In D. Ihde & E. Selinger (Eds.), *Chasing technoscience: Matrix for reality* (pp. 131–144). Bloomington: Indiana University Press.

Jameson, F. (1984). Postmodernism, or the cultural logic of late capitalism. *New Left Review, 146*(1), 53–92.

Kirby, V. (1997). *Telling flesh: The substance of the corporeal.* New York/London: Routledge.

Kroker, A. & Cooke, D. (1988). *The postmodern scene.* Montreal: New World Perspectives.

Kroker, A., & Weinstein, M. (1994). The Hyper-texted body, or Nietzsche gets a modem. *CTheory.* http://www.ctheory.net/articles.aspx?id=144. Accessed 13 Jun 2013.

Lacan, J. (1977). *Ecrits: A selection* (trans: Sheridan, A.). London: Tavistock.

Lakoff, G., & Johnson, M. (1999). *Philosophy in the flesh: The embodied mind and its challenge to Western thought.* New York: Basic Books.

Latour, B. (1988). *The Pasteurization of France.* Cambride, MA: Harvard University Press.

Latour, B. (1992). Where are the missing masses? Sociology of a few mundane artifacts. In W. E. Bijker & J. Law (Eds.), *Shaping technology/building society: studies in sociotechnological change* (pp. 225–259). Cambridge, MA: MIT Press.

Latour, B. (1999). *Pandora's hope: Essays on the reality of science studies.* Cambridge, MA: Harvard University Press.

Law, J. (2009). Actor-network theory and material semiotics. In B. S. Turner (Ed.), *The new Blackwell companion to social theory* (pp. 141–158). Oxford: Blackwell.

Levinas, E. (1969). *Totality and infinity: An essay on exteriority* (trans: Lingis, A.). Pittsburgh: Duquesne University Press.

Levinas, E. (1981). *Otherwise than being or beyond essence* (trans: Lingis, A.). Boston: Marinus Nijhoff Publishers.

Lévi-Strauss, C. (1963). *Structural anthropology* (trans: Jacobson, C. & Grundfest Schoepf, B.). New York: Basic Books.

Levy, N. (2004). Foucault as virtue ethicist. *Foucault Studies, 1*, 20–31.

Lingis, A. (1994). *Foreign bodies.* NewYork: Routledge.

McMaster, T., & Wastell, D. (2005). The agency of hybrids: Overcoming the symmetrophobic block. *Scandinavian Journal of Information Systems, 17*(1), 1–8.

Merleau-Ponty, M. (1962). *Phenomenology of perception* (trans: Smith, C.). London: Routledge & Kegan Paul.

Parisi, L. (2004). *Abstract sex: Philosophy, bio-technology, and the mutation of desire.* London: Continuum Press.

Parisi, L., & Terranova, T. (2000). Heat-death, emergence, and control in genetic engineering and artificial life. *CTheory,* 1–2. Available at http://www.ctheory.net/articles.aspx?id=127. Accessed 22 August 2013.

Pickering, A. (1995). *The mangle of practice: Time, agency and science.* Chicago: University of Chicago Press.

Pickering, A. (2003). Interview with Andrew Pickering. In D. Ihde & E. Selinger (Eds.), *Chasing technoscience. Matrix of materiality* (pp. 83–95). Indiana: Indiana University Press.

Pickering, A. (2005). Asian Eels and global warming: A posthuman perspective on society and the environment. *Ethics & The Environment, 10*(2), 29–43.

Plant, S. (1995). The future looms: Weaving women and cybernetics. In M. Featherstone & R. Burrows (Eds.), *Cyberspace/cyberbodies/cyberpunk: Cultures of technological embodiment* (pp. 45–64). London: Sage.

Poster, M. (1990). *The mode of information*. Oxford: Polity Press.

Poster, M. (2004). The information empire. *Comparative Literature Studies, 41*(3), 317–334.

Rose, N. (2007). *The politics of life itself: Biomedicine, power, and subjectivity in the twenty-first century*. Princeton: Princeton University Press.

Sagan, D. (1992). Metametazoa: Biology and multiplicity. In J. Crary & S. Kwinter (Eds.), *Incorporations* (pp. 362–385). New York: Zone Books.

Sobchack, V. (1995). Beating the meat/surviving the text, or how to get out of this century alive. In M. Featherstone & R. Burrows (Eds.), *Cyberspace/cyberbodies/cyberpunk: Cultures of technological embodiment* (pp. 205–214). London: Sage.

Stelarc. (1997). Parasite visions: Alternate, intimate and involuntary experiences. *Art and Design: Sci-Fi Aesthetics, 12*(9/10), 66–69.

Stone, S. (1995). Split subjects, not atoms; or how I fell in love with my prosthesis. In C. H. Gray (Ed.), *The cyborg handbook* (pp. 393–406). New York: Routledge.

Taylor, M. C. (2001). *The moment of complexity: Emerging network culture*. Chicago/London: University of Chicago Press.

Toynbee, A. (1963). *A study of history*. New York: Oxford University Press.

Turkle, S. (1995). *Life on the screen: Identity in the age of the internet*. New York: Simon & Schuster.

Verbeek, P.-P. (2007). Beyond the human eye: Technological mediation and posthuman visions. In P. Kockelkoren (Ed.), *Mediated visions*. Rotterdam: Veenman Publishers/ArtEZ Press.

Verbeek, P. P. (2008a). Cultivating humanity: Toward a non-humanist ethics of technology. In E. Selinger, J. B. Olsen, & S. Riis (Eds.), *New waves in philosophy of technology* (pp. 241–263). Hampshire: Palgrave MacMillan.

Verbeek, P.-P. (2008b). Obstetric ultrasound and the technological mediation of morality: A post-phenomenological analysis. *Human Studies, 31*(1), 11–26.

Verbeek, P.-P. (2011). *Moralizing technology: Understanding and designing the morality of things*. Chicago: Chicago University Press.

Winner, L. (1980). Do artifacts have politics? *Daedalus, 109*, 121–136.

Zylinska, J. (2002). The future… is monstrous: Prosthetics as ethics. In J. Zylinska (Ed.), *The cyborg experiments: The extensions of the body in the media age* (pp. 214–236). London/New York: Continuum.

Chapter 7
Technologically Produced Nature: Nature Beyond Schizophrenia and Paranoia

Abstract Using the "schizoanalytic" framework of Deleuze and Guattari, this chapter takes a closer look at radical posthumanism by focusing on how this approach analyzes assisted reproductive technologies (ARTs) and their implications for the concept of nature. In radical posthumanist readings, the "schizophrenic" tendency of assisted reproduction to deconstruct the concept of nature is seen as constantly coming up against and being captured by legislative and discursive strategies that "re-naturalize" nature. I argue that this re-naturalization, or reterritorialization, implies a new flexibility with which notions like "biogenetic relatedness", "nature" and "parenthood" are being employed by users of these technologies. And furthermore, that this flexibility indicates new conceptualizations of fundamental categories that are not being adequately accounted for by radical posthumanist discourse, in which deconstructive and disciplining trends coexist and intermingle quite peacefully.

The use of the schizoanalytic framework highlights the importance of radical posthumanism for understanding the shuffling around of foundational terms that is characteristic of emerging biotechnologies. But it also offers a means of pushing these analyses further: in the schizoanalytic framework, reterritorialization is never "just" the reconstitution of a system of meaning that has been unsettled by schizophrenic energies, but always the production of a new one in which deterritorialized elements connect in different ways. This conceptual framework can then complement the notion of technological mediation as developed in the end of the last chapter, i.e., within a context of an ethics of care of the self, to complete the basis of a mediated posthumanist perspective.

Keywords Assisted reproductive technologies • Schizoanalysis • Reterritorialization • Nature • Re-naturalization

T. Sharon, *Human Nature in an Age of Biotechnology: The Case for Mediated Posthumanism*, Philosophy of Engineering and Technology 14, DOI 10.1007/978-94-007-7554-1_7, © Springer Science+Business Media Dordrecht 2014

The relentless debate on emerging bio- and enhancement technologies as it plays out between dystopic and liberal posthumanism is framed as a tension of for and against the widespread use of the technologies.[1] Non-humanist types of posthumanism completely shift this debate, by taking as their starting point a different notion of the human or the subject than the one that is at stake in dystopic and liberal posthumanist discourse. Radical posthumanism, for which the radical critique of modern subjectivity is an essential aspect – more so than for methodological posthumanism, for which this critique is much more implicit – offers interesting directions for conceptualizing an ontology of posthuman subjectivity. In general, foundational categories such as the human, but also nature, are a more important focus of radical posthumanism than methodological posthumanism, and its deconstructive readings of emerging biotechnologies are timely and productive. But radical posthumanism, while moving beyond the polarized framework of dystopic and liberal posthumanism, erects what appears to be another polarized tension, informed by the deconstructive or "postmodern" versus the disciplinary or "modern" tendencies of emerging biotechnologies. The framework of "schizoanalysis" developed by Deleuze and Guattari (1977, 1987) is helpful for articulating this tension as it emerges from radical posthumanist analyses of new technologies. In this schema, emerging biotechnologies embody a great "schizophrenic" potential to challenge and undermine – to deterritorialize – traditional, essentialist understandings of nature, the human and subjectivity. But these technologies also often comprise a "paranoid" tendency, that aims to capture and rechannel this subversive potential – to reterritorialize it – back onto fixed, conventional understandings of nature and the human, thus reinstating rather than invalidating humanist narratives.

This chapter takes a closer look at radical posthumanism by focusing on how this approach analyzes assisted reproductive technologies and their implications for the concept of nature, using the framework of schizoanalysis. In radical posthumanist readings, assisted reproduction is seen as having a disruptive capacity to sever the link between biology, reproduction and traditional understandings of the family. But this schizophrenic potential is also seen as constantly coming up against legislative efforts and discursive strategies that emphasize the importance of a "natural", biological continuity between parents and offspring, that end up *re-naturalizing* nature. I argue that this re-naturalization, or reterritorialization, implies a new flexibility with which notions like "biogenetic relatedness", "nature" and parenthood" are being employed by users of these technologies. And furthermore, that this flexibility indicates new conceptualizations of fundamental categories that are not being adequately accounted for by radical posthumanist discourse, in which deconstructive and disciplining trends coexist and intermingle quite peacefully.

My use of the schizoanalytic framework is thus dual. First, it sheds light on the type of analyses undertaken by radical posthumanist discourse, highlighting the importance of this approach for understanding the shuffling around of foundational terms that is characteristic of emerging biotechnologies. But secondly, it also offers a means of

---

[1]This chapter is inspired by the article "A Schizoanalysis of Emerging Biotechnologies: Renaturalized Nature, the Disclosed Secret of Life, and Technologically Authentic Selfhood" (2011). *Configurations, 19*(3), 431–460.

pushing these analyses further: in the schizoanalytic framework, reterritorialization is never "just" the reconstitution of a system of meaning that has been unsettled by schizophrenic energies, but always the production of a new one in which deterritorialized elements connect in different ways. Various narratives and alternative ontologies do not cancel each other out in these contexts but are layered onto one another in novel reterritorializations that users often draw upon in strategic ways. The new meanings and uses of foundational terms that seem to crystallize, effortlessly, across the posthumanist landscape can thus be accounted for. The use of the schizoanalytic framework can also contribute to a shift in the discussion, insofar as the important question becomes, not how and if deterritorialized elements are being reterritorialized, but if the reterritorializations that are taking place are positive or negative ones. This conceptual framework can then complement the notion of technological mediation as developed in the end of the last chapter, i.e., within a context of an ethics of care of the self, to complete the basis of a mediated posthumanist perspective.

7.1 Biotechnology and Schizophrenia

In the two volumes of *Capitalism and Schizophrenia* (1977, 1987) Deleuze and Guattari distinguish the cycle of de- and reterritorialization as a process by which sets of relations, concepts, or practices are severed or freed from a "territory" (as any kind of system, fixed meaning or context), and then resituated, recontextualized and brought into new relations within a new system or *assemblage*. A simple example of this process is the transformation of a tree branch into a club: the branch is detached (deterritorialized) not only from its physical territory, the tree, but also from its original function, the capturing of sunlight for the tree. It is then resituated in a new territory (reterritorialized), taking on a new function as a weapon (Deleuze and Guattari 1994). The club is thus a deterritorialized branch. A more complex and important example for Deleuze and Guattari is the way that capitalism transforms products into commodities, deterritorializing labor-power by freeing it from specific means of production and reterritorializing it as wages.

The main effect of this process of deterritorialization, while this may not be obvious in terms of tree branches, is that it liberates desire, what Deleuze and Guattari view as the positive and dynamic energy that is the primary reality of subjective and social being.[2] This *schizophrenic* tendency, as the authors call it,

---

[2]Deleuze and Guattari posit desire as the constant production of affective and libidinal energy produced by the unconscious. Their definition runs counter to traditional notions of desire, at least since Freud, that understand it in terms of "lack," in terms of seeking to acquire something that is absent. Against this negative conceptualization, they propose a conception of positive desire that is both real and productive insofar as it establishes real relations with objects and concepts. Desire here is experimental and inventive and does not follow a goal or a direction. It is thus never desire *for* an object; it is in and for itself. Their concept of desire is thus closer to Nietzsche's "will to power" and Foucault's conception of productive power than to Freud's depiction of desire as a drive. This positive construal of desire, however, does not mean that desire necessarily produces positive social formations, since desire can also be made to desire its own repression (1977: 31); rather, it is a case of how desire is arranged and assembled in specific social formations.

once unleashed, comes up against a *paranoid* counter-tendency, a reactionary dynamic that seeks to absorb and recode schizophrenic energies by reterritorializing them back onto transcendental signifiers, social organization and normalizing institutions. Schizophrenia deterritorializes, decomposing value systems, individual and collective identities, freeing desire from socially restricting forces; it is an intrinsically emancipatory process.[3] Paranoia seeks to recompose these, fixing desire in socially accepted representations that regulate which connections desire can and cannot make. This attempt to channel and control schizophrenia, to tame desire's revolutionary and subversive nature, is characteristic of all societies according to Deleuze and Guattari, but capitalism has a special relationship to schizophrenia. The capitalist mode of economic production gives rise to a profusion of decoded flows (in Marx's words, "All that is solid melts into air"), this is to say that it promotes rather than explicitly blocks schizophrenia. Nevertheless, schizophrenic processes, even as they are unleashed by the capitalist machine, are always a threat to its stability, and act as the threshold limit to which it is constantly drawn and from which it persistently pulls back.

7.1.1   The Schizophrenic Pole

Deleuze and Guattari's schizoanalytic terminology lends itself particularly well to radical posthumanist readings of emerging biotechnologies, even though it is not commonly used among radical posthumanist theorists.[4] The deterritorializing capacity of emerging biotechnologies has been an important focus of radical posthumanist scholarship and its theorization of human-technology relations in light of the ongoing crisis of humanism. As I have argued, in this approach advanced technologies are contributing to a breakdown of taken for granted boundaries between the natural and the technological that has rendered the classical humanist framework obsolete. The hybrid entities that technologies produce, it is argued, call into

---

[3] Schizophrenia plays a central role in schizoanalysis, because the breaks and flows that can be observed at the schizoid level escape the Oedipal structure and offer an avenue into our social and historical unconscious. But this is not a celebration of the psychiatric disease of schizophrenia or of the schizophrenic as a clinical patient. Deleuze and Guattari differentiate between schizophrenia as a debilitating psychiatric diagnosis and schizophrenia as a process, as an alternative model of desire-production that provides a clue into its workings. It is when schizophrenia as a process succumbs to repression, when the process of desire-production is interrupted or blocked, that the schizophrenic as a clinical patient is generated:

> The schizophrenics in hospitals are people who've tried to do something and failed, cracked up. We're not saying revolutionaries are schizophrenics. We're saying there's a schizoid process, of decoding and deterritorializing, which only revolutionary activity can stop turning into the production of schizophrenia (Deleuze 1995: 23–24).

[4] An important exception is Rosi Braidotti, who has, more than any other theorist, applied Deleuze's and Deleuze and Guattari's works to feminist readings of biotechnologies. See especially her *Nomadic Subjects* (1994), *Metamorphoses* (2002) and *Transpositions* (2006b). See also Camilla Griggers (1997) and Luciana Parisi (2004).

question the very notion of original, foundational categories, of the givenness of nature, the unity of the organism, and the ontological purity of the human.

In Deleuze and Guattari's schizoanalysis, schizophrenia is a positive tendency, the affirmative potential for freedom, revolution and unbridled creativity. The primary aim of schizoanalysis is to take this preferable tendency to its limits, to push through the limits capitalist paranoia imposes on it. Radical posthumanists, as we have seen, also view the schizophrenic and deterritorializing potential inherent in new technologies as a positive tendency. For radical posthumanists, the problematization and destabilization of the idea of nature as something that is stable and pure – along with the host of other categories for which it has served as a foundation – is something that should be celebrated insofar as appeals to nature have historically been used to legitimate social and sexual hierarchies and norms of human conduct. In this context, new technologies are reconfigured as strategies of resistance against the metanarratives of the Enlightenment and its complicity in colonialist, patriarchal and capitalist structures, that threaten to destabilize the modernist project itself. It is instructive to quote Katherine Hayles and Donna Haraway at length here. Thus at the end of *How We Became Posthuman*, Hayles writes:

> If … there is a relation among the desire for mastery, an objectivist account of science, and the imperialist project of subduing nature, then the posthuman offers resources for the construction of another kind of account. In this account, emergence replaces teleology; reflexive epistemology replaces objectivism; distributed cognition replaces autonomous will; embodiment replaces a body seen as a support system for the mind; and dynamic partnership between humans and intelligent machines replaces the liberal humanist subject's manifest destiny to dominate and control nature. (1999: 288)

And Haraway in *Modest Witness*:

> Nature and Society, animal and man, machine and organism: The terms collapse into each other. The great divide between Man and Nature, and its gendered corollary and colonial racial melodrama, that founded the story of modernity has been breached. The promises of progress, control, reason, instrumental rationality – all the promises seem to have been broken in the children. Man hardly was imagined before he lost his place; nature was barely tamed before she took her revenge; the empire was barely consolidated before it struck back. The action in technoscience mixes up all the actors; miscegenation between and among humans and nonhumans is the norm. The family is a mess. There is hardly a bell curve in sight. Racial purity, purity of all kinds, the great white hope of heliocentric enlightenment for a truly autochthonous Europe, the self-birthing dream of Man, the ultimate control of natural others for the good of the one – all dashed by a bastard mouse [the OncoMouse] and a matched set of unmanly, fictional humans. I find all this to be edifying. Maybe in these warped conditions, a more culturally and historically alert, reliable, scientific knowledge can emerge… OncoMouse and the FemaleMan seem to be co-conspirators in the moral and intellectual terrorism that has been loosed on natural foundations and self-confident rationality. (1997: 120–121)

For these theorists, the deterritorializing potential of new biotechnologies is inherently political, it contributes to the crumbling of the "humanist barricades in the rising tides of posthumanity" (Simon 2003: 2), and supplements the promise of the postmodern project with a powerful technological impetus.

7.1.2 The Paranoid Pole

The schizophrenic potential of new technologies, however, as in Deleuze and Guattari's account of capitalism, is countered by paranoid tendencies that seek to absorb and contain it. These linger in the form of a return of master narratives of technological mastery and scientific progress, discourses of genetic determinism and essentialism, and the continuation or reinforcement of normalizing and conservative trends, all of which contribute to a reinstatement of "modernist" categories rather than their invalidation. In the works of radical posthumanists, this is often presented as a potential/praxis dichotomy: biotechnologies present a schizophrenic *potential* to overcome the essentialisms and binaries of modernity, but in *practice*, the uses these technologies are put to restore foundational categories that are once more used to normalize and discipline. Hayles' thesis, for example, throughout *How We Became Posthuman*, is that her version of embodied posthumanism is an alternative to the prevalent version that has taken shape since the 1950s, epitomized by Hans Moravec's vision of uploaded human consciousness. This kind of posthumanism, as she argues, much like the early cybernetic movement, shuns the problematic implications that the cybernetic paradigm has for the liberal humanist model of autonomous subjectivity, and succeeds in recuperating the posthuman back onto the Cartesian mind/body split. She argues instead for an alternative, embodied posthumanism that can potentially be born out of the interface between human bodies and computer-based technologies. But this is a *potential*, one that was suppressed by the early cyberneticians as it is by the dominant form that the posthuman takes on today in transhumanist accounts. It is a vision that needs to be salvaged:

> Of course, this is not necessarily what the posthuman *will* mean – only what it *can* mean if certain strands among its complex seriations are highlighted and combined to create a vision of the human that uses the posthuman as leverage to avoid reinscribing, and thus repeating, some of the mistakes of the past. (288)

Haraway too, continues from the quote above: "It remains to be seen whether the rush-hour traffic across the boundaries of nature and culture in genome discourse constitutes a case of fluid practice or a particularly grave case of hardening the categories in technoscience" (1997: 149). And she concludes that, "For all their inventiveness in making fabulous natural/cultural hybrids that circulate fluidly in vast networks, many actants ... seem 'to be suffering from an *advanced case of hardening of the categories*'" (1997: 169, emphasis added).

This pessimistic inference is shared by many radical posthumanists.[5] For example, Jill Didur, writing about genetically modified foods, argues that,

> Despite the rhetoric of hybridity and constructivism that characterizes ... claims about the impact of these new technologies in society, their ownership, implementation, and regulation are haunted by an Enlightenment subject that presupposes knowledge as disembodied and humans as autonomous and unified agents, and ultimately *reinscribes relations of power* along colonial lines. (2003: 100, emphasis added)

---

[5] I quote these theorists at some length in order to highlight that there is a clear sense in which paranoid trends override the schizophrenic potential in these writings.

Or, Anne Balsamo, in her work about the gendered body in contemporary technoscience:

> As is often the case when seemingly stable boundaries are displaced by technological inno- vation (human/artificial, life/death, nature/culture), *other boundaries are more vigilantly guarded.* Indeed, the gendered boundary between male and female is one border that remains heavily guarded despite new technologized ways to rewrite the physical body in the flesh. So it appears that while the body has been recoded within discourses of biotechnol- ogy and medicine as belonging to an order of culture rather than of nature, gender remains a naturalized marker of human identity. (1996: 9, emphasis added)

Teresa Heffernan, in her examination of the public discourse around the ethics of developing transgenetic organisms, writes:

> Pig valves in transplant patients or tissues grown with the aid of a cow egg or hamster eggs fertilized with human sperm to test fertility or pigs spliced with human genes are all accept- able hybrids in the construction of the new post-Enlightenment body of science because, in the process of the assimilation of the "non-human", the hierarchical divide between it and humanity is sustained. The owning, controlling, patenting, and manipulation of what is understood as nature (as excluding humanity but in its service) is *left unchallenged*; the boundary between the monster and the human *is secured*; the notion of the human as a well- defined category distinct and autonomous from the nonhuman is *left unquestioned* even as the production of the human is enabled by the nonhuman. (2003: 128, emphasis added)

Guattari, writing much later than *Anti-Oedipus* on information technology:

> The burning question, then, becomes this: Why have the immense processual potentials brought forth by the revolutions in information processing, telematics, robotics, office automation, biotechnology and so on up to now led only to a *monstrous reinforcement* of earlier systems of alienation, an oppressive mass-media culture and an infantilizing poli- tics of consensus? What would make it possible for them finally to usher in a postmedia era, to disconnect themselves from segregative capitalist values and to give free rein to the first stirrings, visible today, of a revolution in intelligence, sensitivity and creativity? (1992: 29–30, emphasis added)

And Braidotti writes even more straightforwardly, and using Deleuzian terminol- ogy, that "the potentially innovative, de-territorializing impact of the new technolo- gies is *hampered and turned down* by the reassertion of the gravitational pull of old and established values". (2006b: 2, emphasis added)

The paranoid capture that these theorists identify in many biotechnological con- texts begs the question as to why the showdown between schizophrenic and para- noid forces is ultimately resolved by the appropriation of the former by the latter and its grafting back onto the "modern project". Accusing fingers are usually pointed at capitalism in its various manifestations. For Haraway, the "dark side" of posthumanity, the "informatics of domination", is a product of global capitalism and militarism. For Braidotti, it is a result of globalized capitalism's drive towards commodification and profit-oriented differences. If the capitalist mode of produc- tion multiplies differences, these, she argues, are simultaneously redefined as a matter of lifestyle choices, subject to the logic of consumer culture, and consequently grounded in new cultural essentialisms.

The problem of how advanced capitalism succeeds in co-opting subversion – of transforming counter-cultures into mainstream, of commodifying difference and of thriving on multiplicity and fluidity – has preoccupied a number of theorists of the postmodern, from Jameson (1984) through Hardt and Negri (2000). In this sense the paranoid capture of the schizophrenic potential of emerging biotechnologies can be seen as the playing out of this larger phenomenon on a smaller scale, which also preoccupies radical posthumanists. On this "smaller scale", radical posthumanists also mull over the worrisome complicity between critical theory and capitalist ethos: the unfortunate adoption by private biotech companies and public scientific discourse of a vocabulary of heterogeneity, flexibility, boundary transgression and difference, a vocabulary over which postmodern theory once had sole custody. Thus in the introduction to a special issue of *Cultural Critique* (2003) on the posthuman, the need to disentangle the "critical potential of hybrid subjectivity" from the production of material hybrids in the scientific realm is upheld as a pressing task of the growing discipline of "critical" (or radical) posthumanism.

This dynamic, by which paranoid forces seem to prevail, also runs throughout Deleuze and Guattari's schizoanalysis, where schizophrenic energies are usually absorbed and rechanneled by the existing system which ultimately blocks their revolutionary potential. On their account, "there is no deterritorialization of the flows of schizophrenic desire that is not accompanied by global or local reterritorialization" (1977: 347). But the processes of de- and reterritorialization need to be understood as a *continuum* more than a simple binary, since every territory already includes "vectors of deterritorialization" and every deterritorialization is already "inseparable from correlative reterritorializations" (1987: 509). Every system or assemblage thus continuously oscillates between de- and reterritorializations, a haunting potential that looms within each's other. Yet the incessant nature of this oscillation implies that it is not just a case of positive deterritorializations and negative reterritorializations, but a question of positive and negative forms each of de- and reterritorializations, that depend on the nature of the new relations deterritorialized elements come into – which can be either productive connections, or obstructive conjugations – within the new system (1987: 220).[6] This point is extremely important, insofar as this complexity of the movements of de- and reterritorialization provides a means of accounting for the very rich transformations that notions of nature, life and selfhood are undergoing in light of emerging biotechnologies, that express more than just a paranoid capture of schizophrenic, deconstructive energies.

This framework allows us to move away from a simple and evident dichotomy of liberating versus repressive trends in the ways new biotechnologies destabilize foundational categories, to see how these interact, connect and co-evolve in novel understandings that are more than the resuscitation of their former selves. This is an insistence on the irreducibility of mixtures, a move beyond hybridity, by which reterritorialization signifies at the least a variation of an older system of meaning if

---

[6]They discuss this mainly in terms of "lines of flight". In *Dialogues* Deleuze claims: "This is why the question of schizonalaysis or pragmatics, micro-politics itself, never consists in interpreting, but merely in asking what are your lines, individual or group, and what are the dangers on each" (1987).

not the constitution of a new one. In this context, the important question can become not "are deterritorialized elements being reterritorialized?", but "are the reterritorializations that are taking place positive ones?" – though it remains to define how this judgment can be made. The schizoanalysis of emerging biotechnologies I propose here as a supplement to radical posthumanist analyses thus attempts to account for new phenomena, new understandings of foundational categories in light of the interaction of schizophrenic drives and paranoid pulls that are emerging in the current biomedical landscape.

7.2 A Schizoanalysis of Assisted Reproductive Technologies

7.2.1 The De- and Re-Naturalization Nature

The rapid development of assisted reproductive technologies (ARTs) over the past 30 years – including procedures such as in vitro fertilization (IVF), artificial insemination, donor insemination, preimplantation genetic diagnosis (PGD), hormone treatment, surrogacy and cryopreservation – has become a sign of the increasing prominence of biotechnologies in some of the most basic areas of human existence. As highly technologized forms of intervening in nature, ARTs are often presented in radical posthumanist discourse as providing grounds for unique forms of schizophrenic and postmodern experimentation in social and kinship relations. Theorists emphasize the disruptive potential of ARTs, arguing that by severing the link between heterosexual and biological reproduction, these technologies make it possible to subvert, fragment and transform conventional meanings of "nature", "gender", "reproduction", "motherhood" and "the family" in radical ways. At the same time, these technologies are also seen as perpetuating and reinstating forms of paranoid and modern biopolitical control, via normative assumptions, discursive strategies and legislative efforts that work to identify legitimate uses and users of such technologies while simultaneously affirming conventional understandings of "nature", "gender", "motherhood" and "the family".

The schizophrenic effect of reproductive technoscience has had its greatest impact on kinship ties and notions of the family by contributing towards a greater fluidity in kinship patterns and pluralizing notions of relatedness. The use of technologies and third parties in the process of procreation has led to a destabilization of the biological aspect of parenthood. As Marilyn Strathern has noted, ARTs have created "a new convention, the distinction between social and biological parenting, out of an old one, kinship and the social construction of natural facts" (1992: 27–28). Strathern argues that if kinship, a set of social relations, was previously seen as being rooted in the natural facts of biological reproduction, then ARTs have a de-naturalizing effect on kinship models, thus blurring the nature/culture dichotomy. Core notions of kinship and conventional family units are thus unsettled by ARTs, as a multiplicity of forms of quasi-, semi- or pseudo-biological forms of parenthood are created (Franklin and Ragone 1998). With the introduction of surrogate motherhood, for example, the concept of "motherhood" is deconstructed into a variety of possible functions, of genetic, birth, adoptive and surrogate maternities (Inhorn and Birenbaum-Carmeli 2008). Similarly, donor insemination can disrupt ideals of manhood and fatherhood, subverting models of heteropatriarchal kinship. These developments destabilize any axiomatic link between biology and parenthood, and indicate ways in which supposedly "natural" categories already conceal multiple meanings. Furthermore, some argue, they might point to the loss of an essentialized definition of womanhood, or at least of motherhood, indicating the decline of one rigid or fixed way of experiencing motherhood in favor of an increased freedom for women.[7] The capacity for reproduction to transcend individual bodies via ARTs increasingly means that the achievement of biological parenthood is not limited by one's age, marital status, sexual orientation or natural fecundity, thus allowing for renewed understandings of who can become a parent and under what circumstances.

Similarly, ARTs convey a schizophrenic potential to unsettle conventionally exclusive definitions of the "family" as heterosexual and nuclear, by facilitating the creation of non-traditional family forms. ARTs can expand categories of parenthood beyond reproduction by strictly heterosexual intercourse to gay, lesbian and transgender couples, as well as single parents. Reproduction in same-sex unions currently focuses on donor insemination and gestational surrogacy, but recent intersections between assisted reproduction and human genomics hold the promise of countless reproductive possibilities, from cloning and gene-splicing technologies to attempts to transform male stem cells into eggs and female stem cells into sperm that might allow same-sex reproduction.[8] This "queering of reproduction" (Mamo 2007) poses an obvious challenge to patriarchal reproductive hegemony, as many legislators and social commentators have decried (Halberstam and Livingstone 1995).[9] Increasingly, conventional sexual relations and related family structures are no longer necessary for reproduction, and by completing the dissociation between sex and reproduction that was begun with the introduction of the contraceptive pill, ARTs offer a broad array of partnerings and possibilities for reproduction. As access to parenthood is widened, the idea of natural conception becomes increasingly problematic. Nature as a foundational concept is twice unsettled here: first as natural biological reproduction, and second, as what a natural reproductive unit is.

---

[7]This is the same liberatory component identified by some early feminist scholarship on reproductive technologies such as Shulamit Firestone's (1970) radical advocacy of their use to free women from the "tyranny of reproduction".

[8]Eskridge and Stein (1998) advocate what they call "queer cloning" as a technology that offers the possibility of completing the perfect segregation of sex and reproduction, and the "next logical step in queer people's formation of families of choice" (109).

[9]As attests the *Daily Mail*'s review on male egg and female sperm techniques, titled "Death of the Father". See Fiona Macrae (2008).

But the schizophrenic potential inherent in ARTs is countered by several paranoid tendencies. Many feminist scholars have argued that far from liberating women from the burdens of childbirth, the technologization and medicalization of conception has transformed women's bodies into sites for increasing medical intervention and control (Balsamo 1996; Braidotti 1994; Duden 1993; Hartouni 1991; Stabile 1994; Shildrick 1997). In this sense there has been a literalization of the modernist Baconian metaphor of the pursuit of scientific knowledge as the domination of the female body of nature (Jacobus et al. 1990). A significant concern among these theorists is the degree to which ARTs replicate traditional, patriarchal, ideologies of motherhood and femininity that reduce women to their reproductive bodies. Thus, it is suggested that ARTs lead to an institutionalization of scientifically managed reproduction, that processes of isolation and visualization contribute to a fragmentation of the female body that leads to an erasure of women's subjectivity, and that they constitute the female reproductive body as an entity that needs to be monitored. Stripped of bodily sovereignty, the reproductive female body in such narratives is understood both as a factory of potential persons and as potentially criminal;[10] a body that needs to be incorporated into systems of normative surveillance and necessitates discipline.

The schizophrenic, "post-biological" potential of ARTs is furthermore seen as being captured by paranoid attempts to re-establish the link between reproduction, heterosexuality and the nuclear family. Medical, legal and moral rhetoric, this argument goes, contribute to maintaining and institutionalizing a conservative network of reproduction, thwarting the proliferation of new opportunities for parenthood by selectivity that is based on social suitability. While explicit legislative efforts to restrict access to fertility clinics are usually not successful, fertility laws often incorporate value judgments and biases that channel the demand for ARTs in conservative directions, indicating, again, the highly subversive potential of these technologies. Most commonly, this can be seen in cases where single women and lesbians are denied access to fertility services. Thus various "liberal" countries including Austria, France, Denmark, Ireland and Sweden, formally restrict access to donor insemination and IVF to married couples or, in some cases, to heterosexual couples in stable de facto relationships (Bryld 2001; Pattinson 2003).

In the UK, for example, the Human Fertilization and Embryology (HFE) Act of 1990 states that, "A woman shall not be provided with treatment services unless account has been taken of the welfare of any child who may be born as a result of the treatment (including the need of that child for a father)" (s.13(5)). This stipulation has enabled some clinics to deny same-sex couples and single parents IVF treatment

---

[10] Stabile (1994) writes that the maternal body is discursively transformed from "a benevolent, maternal environment into an inhospitable wasteland, at war with the 'innocent person' within" (70). See Petchesky (1987) for a discussion on the politics of fetal imaging and Newman (1996) for a history of fetal images from the sixteenth century to today. These authors argue that the fetus was "born" as a distinct entity through rhetorical and visual techniques that severed it from the maternal body and invested it with "life" and subjectivity.

through statutory interpretation.[11] The fact that such restrictions are often justified as having the child's best interest in mind, based on the presumption that children need two parents living at home, and that one of these must be the child's father, is interpreted by critics as an expression of the cultural anxiety caused by the idea of same-sex families and the need to safeguard heterosexual normativity.[12] The Australian government has expressed similar intents on banning single women and lesbians from accessing IVF services. The state of Victoria, for example, imposed such a ban through the Infertility Treatment Act of 1995, which stated that "a woman who undergoes a treatment procedure must be married" (s.8(1)). A court later judged this provision to constitute unlawful discrimination on the grounds of marital status, but, similar to the situation in the UK, while access has been widened to non-married couples, it is restricted to those deemed medically infertile, thus excluding in practice the majority of single women and lesbians from any form of treatment (Liu 2009).[13] In such examples, the distinction between the use of ARTs to treat infertility and its use to resolve a couple's or an individual's "involuntary childlessness" is upheld in a way that echoes the treatment/enhancement dichotomy.

---

[11] Though this criterion was amended with the HFE Act 2008 following threats of legal action by lesbian couples. In the amendment, the "need for supportive parenting" rather than the "need for a father" is stipulated (Blackburn-Starza 2008).

[12] For a discussion on the "child's best interest" in this context, and the lack of empirical evidence suggesting that same-sex parenthood is not in that interest, see Brewaeys et al. (1997); Lycett et al. (2004), Murray (2004) and Walker (2003).

[13] Such restrictions are particularly significant in countries where these infertility services are subsidized by the state. In the US, in contrast, fertility clinics are private and there is little if any regulation. While the profession is largely profit-driven (Spar 2006), clinics are also free to reject individuals seeking treatment, and this has led to efforts on the part of professional and federal bodies to define guidelines to maintain open access. In 2009, the Ethics Committee of the American Society for Reproductive Medicine (ASRM 2009) published a statement that denial of access to fertility services on the basis of marital status or sexual orientation is unjustified on ethical grounds. Also, in August 2008, California's Supreme Court ruled in favor of a lesbian couple who had been denied treatment based on religious objections, and claimed that physicians must offer infertility services to gays and lesbians despite religious objections, or find a colleague in their office who will do so (Dolan 2008).

Another exception is the state of Israel, which offers a unique and fascinating case in this international context. In Israel, all attempts to restrict ART provision have failed in both court and Parliament due to a complex matrix of historical, religious and political motivations including the biblical tenet to "be fruitful and multiply" and the notion of procreation as a means to ensure Jewish continuity in light of the Holocaust and the "demographic threat" (Birenbaum-Carmeli and Dirnfeld 2008; Haelyon 2006; Kahn 2000; Shalev and Goolding 2006). This has led to the implementation of one of the most aggressive and proactive ART policies in the world, where IVF is almost completely subsidized by the state and open to any Israeli woman (including Arab Israelis) irrespective of marital status or sexual orientation, until she has two children with her current partner. Furthermore, contrary to what one might expect in light of the Catholic Church's denunciation of all forms of assisted reproduction and their ban in Catholic countries such as Costa-Rica, Ireland and Italy, rabbinical law in Israel has actually been particularly receptive of these technologies, which are viewed as consistent with religious views of kinship and family formation (Kahn 2002), even and including some of the most controversial forms of ART, like PGD for sibling donor (Hashiloni-Dolev 2007).

Thus, while these technologies expand the range of possible maternal subject positions to "infertile", "postmenopausal" and "lesbian mothers", in addition to those mentioned above, their regulation is seen as drawing those possibilities back into normalized categories of motherhood. Similarly, the reterritorialization of hetero-normativity is echoed in what theorists have noted as the perpetuation of racial purity in assisted reproductive practices. The idea here is that while technologies like donor insemination could be used to subvert the model of the racially unified family, as well as the very idea that race is reducible to biological features, what we find is that the racial characteristics of donors is one of the most important and carefully catalogued criteria presented by sperm banks. Szkupinski Quiroga (2007) makes this point in her article on how race is deployed in biomedical solutions to infertility. She writes,

> Donor insemination can also present a threat to the essentialist notion of whiteness … The "unknown" genes of the sperm donor represent a potential destabilizing threat to the illusory purity of race as evidenced by physical markers. (150)

But, she continues,

> Sperm banks simultaneously manage the subversion of patriarchy and racial purity through the careful cataloguing of donors' physical characteristics, which are then used as a basis for "matching" and choosing the appropriate donor. (150)

Szkupinski Quiroga analyzes cases of women being mistakenly inseminated with sperm from the wrong racial group and the degree of media outcry in response to them as evidence of this.[14]

7.2.2   Rhetorical Resolutions

From a radical posthumanist perspective, the immense schizophrenic potential inherent in technologies of assisted reproduction – as in all biotech and enhancement technologies – is the problematization of the idea of nature as something that is given, stable, pure and superior to the "artificial". But this is not a reiteration of the objection voiced by liberal posthumanists to the dystopic posthumanist argument for the givenness or "sacredness" of nature. The radical posthumanist celebration of these technologies' capacity to undermine the idea of nature proceeds not from the belief that humans can and should manipulate nature at will, but from the understanding that, historically, appeals to nature have been used to legitimate social and sexual hierarchies and norms of human conduct. It is the potential to challenge the foundational character of the category of nature, in light of the understanding that it has been used as a strategy for maintaining boundaries, for political or economic ends, for reifying cultural values, when it really does not refer to any object or category in the world, that radical posthumanist discourse emphasizes.

---

[14] Meanwhile, celebrity adoptions strive to emphasize racial discontinuity as defined by visual markers as much as possible.

In this sense, the dystopic posthumanist critique of bio- and enhancement technologies is interpreted by radical posthumanists as an attempt to hold onto a concept of nature that legitimizes exclusionary practices.

This critique of nature, as with regards to subjectivity, is taken up from earlier poststructuralist critiques that emphasize nature's discursive status, its instability and its lack of any fixed reference. Foucault (1973, 1979), for example, presents the distinction between the "natural" and the "unnatural" (or "perverse") as itself discursively constituted, and the explanatory force of the reference to a common natural foundation is rejected. And Derrida (1976, 1981) inscribes nature in the logic of supplementarity, by which a primary term always makes possible, through the impossibility of its own full presence, its binary opposite: that term which has been expelled in order to constitute it. Nature here is always a construct, created between "absence" and "presence", impossible to pin down yet necessary for the production of meaning. ARTs seem to mirror this conceptual dissolution of the concept of nature and its constructed character in a material sense, by continuously blurring the distinction between the natural and the technological: how can technology be nature's other when notions like reproduction, life, basic elements of what is most natural, are being technologically produced on such a large scale? As Strathern has claimed, "biology under control is no longer 'nature'" (1992: 35). We might expect the very concept of nature to loose all of its foundational authority in this context, if not to become completely evacuated of meaning; a term we no longer have any use for. But the schizoanalytic framing of ARTs undertaken above seems to attest to the contrary, to the grounding of a paranoid *continuity* between reproductive technologies and discourses of natural processes.

In other words, what should emerge as a troubling oxymoron – "technologically produced nature" – is resolved in the framework of reproductive technology by the creation of what seems like an innocent euphemism – "nature's helping hand": ARTs are not understood as alternatives or substitutes to natural conception, but as tools that promote, supplement, correct or improve natural processes that have gone wrong. This is suggested in the very term "*assisted* reproductive technologies", a framing that contributes to a domestication of fears about its unnaturalness.[15] The assistance that ARTs provide implies that nature is unpredictable and imperfect "by nature", something that "naturally" breaks down once in a while and needs a bit of a helping hand to "get back on track". In other words, if these technologies have the potential to change normative expectations about kinship, heterosexuality and the nuclear family, they are often viewed as tools to *restore* a normative state. And despite the plurality of types of families that ARTs have helped usher in, they are often seen as enabling couples to have the "same" kinds of families as "everyone else" (Franklin and Roberts 2006).

In the schizoanalytic dialectic, nature, rendered malleable on the one hand, is re-established as a legitimizing force on the other. Thus, just as it becomes

---

[15] Hartouni (1997) and Franklin and Roberts (2006) discuss how this took place for IVF techniques. Franklin and Roberts explain how anxieties surrounding IVF have now been passed on to the PGD technique.

increasingly difficult to ground nature and uphold the distinction between nature and technology, a greater emphasis seems to be placed on the desire to create a "natural" continuity between parents and offspring, a desire which takes on the feature of a self-evident right in terms of demands for state subsidization of access to ARTs. And just as the notion of parenthood is increasingly severed from a strictly unmediated biological basis, genetic inheritance seems to become privileged over all other forms of kinship ties. A number of practices seem to corroborate this. Many women (and men) are willing to pay a high physical and emotional price in order for fertility treatments to succeed. Furthermore, studies show that adoption, as an alternate, non-medical means for family formation by infertile couples is being marginalized (Storrow 2006). Just as a shroud of secrecy often surrounds the use of donor sperm among heterosexual couples, and "resemblance talk" is used to mask a child's origins (Becker et al. 2005). Elizabeth Sourbut writes:

> The discourse of biological destiny defines infertility within the terms of biological science. Heterosexuality and the nuclear family are presented as natural. Only procreation through heterosexual intercourse within marriage is seen as producing legitimate parenthood. This, of course, is profoundly disrupted by the new reproductive technologies, which not only bypass heterosexual intercourse, but often involve third-party donations of gametes. This contradiction can only be contained by rigid discursive conventions which privilege genetic inheritance over all other forms of parent–child ties. (1996: 236)

The tension between the schizophrenic and paranoid tendencies of ARTs in regards to nature can be identified in other emerging biotechnologies as well. The same dynamic, for example, is at work in the research and development of genetically modified organisms (GMOs). GMOs are enveloped in a rhetoric of nature as originally hybrid and constructivist. Indeed it is the very assumption that species boundaries are permeable that legitimizes from the outset the creation of transgenic breeds. This assumption also underlies the claims made by biotech companies that the genetic modification of crops in a laboratory does not differ in essence from the cross-breeding of plants "in nature" – either in the absence of human intervention or guided by century-old agricultural methods like selective breeding or the controlled pollination of plants. Plant identity thus emerges as always having been the result of a dialogue between nature and culture, whether in or outside the lab. As Haraway (1997: 60) has noted, this schizophrenic potential for transgenic border-crossings poses a serious challenge to the "sanctity of life", since the transferring of genes between species transgresses natural barriers, compromises species integrity and violates the essential and intrinsic quality of nature.

As with ARTs, in this context the genetic modification of crops can be presented as improving nature, an alternative that is superior to nature but that really merely does what nature naturally does, only more safely and efficiently. This underlies the shift in terminology from genetically "engineered" to genetically "modified" organisms (Levidow 1996), which reflects an attempt to resolve the paradox of "technologically produced nature". But the rhetoric of improving, enhancing or even "perfecting" nature in the context of GMOs always relates to the improvement or perfection of nature *for* the sake of humanity – a slippage which re-establishes the hierarchical nature/culture divide and reinforces the modernist narrative of scientific progress as the domination and exploitation of nature as inanimate matter. This is evident in the case of patenting laws: while the possibility of producing new varieties of plants or organisms assumes that nature is essentially hybrid and constructivist, it is the assumption that there is a fundamental difference between nature and culture, or nature and human activity, that provides the basis for patent claims in agribusiness. Without a clear nature/culture divide, the distinction that informs patent laws between the simple "discovery" of plant varieties as opposed to their "invention" would have no foundation; the notion of human authorship of a genetically modified organism implies a consciously acting agent and a passive natural world. As with ARTs, the technology of creating GMOs seems to both unsettle and then reaffirm the legitimizing power of the category nature.

7.2.3 The Strategic Mobilization of Re-Naturalized Nature

As we have seen here, the notion of nature, and its correlate terms in the context of ARTs such as parenthood, reproduction and family, once de-naturalized, are just as soon "re-naturalized" (Franklin et al. 2000). The analyses of ARTs undertaken in radical posthumanist discourse shed light on this important and characteristic process of emerging biotechnologies. But these analyses too often stop at the moment of paranoid "capture" of schizophrenic potentials, without giving a full account of how complex the process of reterritorialization in these contexts is.

In Deleuze and Guattari's schizoanalysis, the movement of reterritorialization does not imply a return to the original territory, but a recombination of deterritorialized elements into new relations in a new or modified system. Renaturalized natural categories, in this sense, are not the same as they were prior to their denaturalization. This is to say that it is not only the case of a paranoid capture of the schizophrenic potential inherent in a malleable or technologically produced nature, as is often presented by many radical posthumanist readings of biotechnologies, but an interaction between these two forces that gives rise to new understandings of these categories that take on a new plasticity and flexibility. In other words, the fact that nature and categories that are linked to naturalness, such as reproduction and parenthood, have not been completely done away with does not necessarily mean that a negative reterritorialization has taken place.

This seems to be an important claim of a growing number of ethnographic works on the real-life experiences of women and men using assisted reproductive techniques in the field of kinship and biomedicine, and their emphasis on the considerable flexibility with which notions like "biogenetic relatedness", "nature", "parenthood", "birth" and "nurture" are being used in these contexts (see for example Franklin and Roberts 2006; Ragone 1994; Thompson 2005; Strathern 2005). These authors have noted the ways in which users of these technologies, as individuals, couples and extended reproductive units including donors and surrogates, "strategically naturalize" technological assistance to conception, a process which allows them to normalize and legitimize practices that might otherwise be seen as deviant.[16]

In her study on surrogate motherhood, for example, Helene Ragone (1994) argues that the tendency to assume that the primary motivation of couples using ARTs is to have a child who is biologically related to at least one of the members of the couple obscures a much more complex state of affairs. The dilemmas raised by the destabilizing power of surrogacy are resolved, she argues, by a number of strategies employed by all participants in the surrogate process that rework categories, namely of motherhood, into intelligible kinds. One strategy is to de-emphasize the role of the adoptive father in order to stave off the looming specter of adultery implicit in his relationship to the surrogate mother. Another, more importantly, consists in bringing to the fore the importance of the social, nurturing role played by the adoptive mother. This aims to resolve the adoptive mother's lack of genetic relatedness to the child, by downplaying the surrogate's genetic contribution and stressing the idea that it is the adoptive mother's desire to have a child that is the origin of, and that which makes possible, the surrogate birth.

Charis Thompson (2005) offers a similar account in her study on gestational surrogacy and egg donation and a telling comparison of both. She describes how couples strategically naturalize their use of the technologies, "foregrounding" or "minimalizing" certain aspects of the technologies in order to construct themselves as the legitimate, "real" parents of the prospective child. Thompson found this to be true and carried out in a similar fashion for technologies that can be seen as performing contrary tasks (though with the overall goal of "making parents"): for a group of couples who used gestational surrogacy and embryos made from their own gametes, as well as for a group who used donor eggs but carried the fetuses themselves. For the first group, the genetic connection, not gestation, was upheld as the most important aspect of their bond to the future child. While in the second group gestation, not genetic continuity, was seen as the most significant aspect of this bond. Biogenetic continuity was either emphasized or downplayed in order to reconfigure the technologies as fundamentally *similar* to "natural" conception. As one of the participants in Franklin and Roberts' (2006: 184) study of preimplantation genetic diagnosis eloquently states regarding "nurture" and the "genetic tie": "they are more or less shades of the same thing".

In these examples, the deterritorialized notions of nature, parenthood, natural conception and birth enter into novel connections that were not possible before the technologies existed. For every normalizing strategy that succeeds, for every re-contextualization of technical components within a new, legitimate understanding of "how babies are made", a reterritorialization has taken place. This process is unceasing and characterized by a remarkable flexibility, an ease with which individuals using these technologies can alternate between different meanings of what

---

[16] While the strategy of "naturalization" is particularly significant in the context of biomedical technologies like ARTs, it might be seen as a common aspect of the integration or domestication of all technological innovations which raise initial fears of "dehumanization" or "excessive human hubris".

natural means. It manifests what Franklin and Roberts (2006) call the "digital" quality of biological identity, in which different, seemingly contrary aspects of conception, identity and kinship, can momentarily be brought into focus or be minimized, while never cancelling each other out.[17] This process is more complex than a mere paranoid capture of the destabilizing effects of ARTs via a clear reinforcement of the boundary lines between nature and culture, or a simply negative movement of reterritorialization that stifles the innovative and liberatory potential inherent in these technologies. This is not to say that conservative strategies that extend old concepts are not part of this dynamic, but that there is much more going on.

We might say that if the notion of nature indeed maintains something essential and fixed in the process of renaturalization, this pertains much more to its *structure*, as a foundational category that normalizes and legitimizes, but that the *content* it receives is constantly changing, and allowing for the inclusion of aspects of being human that could not be in the past. The question then becomes, not if or how the liberatory potential of these technologies has been co-opted, but if the reterritorialization that has taken place is a positive one. Thompson seems to view her analysis in similar terms. In her study, she labels *ontological choreography* the coordination of a "deftly balanced coming together of things that are generally considered parts of different ontological orders (parts of nature, part of the self, part of society)" (2005: 8). We can identify this coming together as a reterritorialization, the different parts being deterritorialized understandings of various concepts. These different parts need to be "coordinated", in Thompson's words, or "connected" in Deleuze and Guattari's, in what Thompson calls "highly staged ways so as to get on with the task at hand: producing parents, children and everything that is needed for their recognition as such". This reterritorialization can be seen as a positive one, according to Thompson, insofar as it leads to "ontological innovation", what we can understand as new ways of making children and making parents for individuals who could not in the past, of bringing together deterritorialized elements, practices and notions so that they find a new coherence that is functional and productive in a specific setting.

7.3   Conclusion

The effects of emerging biotechnologies in radical posthumanist discourse can be seen in terms of the schizoanalytic framework, as having schizophrenic and paranoid tendencies. In the context of assisted reproduction, the schizophrenic drive of new technologies threatens to undermine nature as a stable, foundational category by severing the link between heterosexual and biological reproduction. The paranoid drive, conversely, reinstates nature as a normalizing category, perpetuates forms of biopolitical control and re-establishes the link between reproduction,

---

[17] Franklin and Roberts use the term "digital" to refer to the "switching back and forth" between nature/culture, born/made that is a recurrent theme in kinship theory.

heterosexuality and the nuclear family. Clearly, this is a completely different conceptual framework from which to look at emerging biotechnologies than the humanist one that underlies the dystopic/liberal posthumanist debate, and it makes two important contributions to the discussion on emerging biotechnologies.

First, such analyses shed light on the shuffling around of foundational terms that emerging biotechnologies are contributing to, on their deconstructive and deterritorializing effects, something that none of the other approaches, including methodological posthumanism, are equipped to do. And this is significant, because the "shuffling around" that is caused by the movements of de- and reterritorialization may be precisely what it is that is common to emerging biotechnologies, and what makes them the cause of so much anxiety and hope: the idea that nature, or biology, is both given *and* given to control, that the natural or biological existence of things is determined *and* that it can be intervened upon and remade. Secondly, radical posthumanist analyses nurture a sharp critical appreciation of emerging biotechnologies, keeping an eye, so to speak, on the paranoid tendency that often animates them. This critical tool should be an important aspect of any improved posthumanist perspective as well, and it is taken up by mediated posthumanism. This may seem in contradiction with the claim I made in the previous chapter, that the celebration of hybridity that we find in radical posthumanist discourse is often disconnected from the material practices in which hybridity is being engendered. But the contrary is true. An astute evaluation of the paranoid effects of biotechnologies does not preclude the celebration of their vast schizophrenic potential, and it is precisely the potential/praxis dichotomy that we have seen to be a part of radical posthumanist discourse here, that also underlies the celebration of hybridity and the view of posthuman subjectivity as a form of political resistance. This dichotomy can in a sense be seen as a result of radical posthumanist readings of emerging biotechnologies. To the burning question: "why has the great potential of these technologies not yet been realized, and why have we not yet moved into a post-anthropocentric posthuman episteme?" it offers a clear response: "Because the practices that surround us are still guided by predominantly humanist, disciplinary rationales, by which the paranoid captures and rechannels liberatory schizophrenic potential".

To be sure, there *is* a tension between the deconstructive and destabilizing effects of emerging biotechnologies and their potential to reinforce traditional patterns of power. But these should not be cast in sharp contrast. The analytical tendency to dichotomize the paradox of nature's given and technologically produced character prevents us from seeing how re-naturalized nature is much more than a return to nature prior to its de-naturalization. If its powers of authentification and legitimation are often reinstated, it is also a more malleable and compliant notion of nature, whose boundaries are negotiable, and which allows for the inclusion of many forms of experiences that would not be considered natural in the past. If this flexibility has always been a defining characteristic of the notion of nature, this is, I believe, more evident for the notion of nature in our technological culture. This is where the schizoanalytic framework offers an important contribution, as I have argued, insofar as reterritorialization is never just the reconstitution of a system of meaning that has been unsettled by schizophrenic energies but always the production of a new one in which deterritorialized elements connect in different ways. This type of insistence on the irreducibility of mixtures is necessary for grasping many of the phenomena that speckle our biotechnologized landscape, in which seemingly contradictory trends and concepts interact and coexist with an often surprising ease. As in the example of nature in the context of ARTs, there is an overlapping of essentialized and de-essentialized understandings of natural notions, of the determinedness and the transformativity of biology, a "digital" way in which these are at times emphasized and at times down played but never completely replace one another. In the next chapter, I will discuss how this kind of understanding, integrated in a mediated posthumanist approach, can help counter claims of a general "geneticization" or "biological essentialism".

I also claimed that this kind of understanding, or perspective, shifts the debate in a different direction, opening it to the question of what makes for a positive reterritorialization. The criteria for this kind of normative evaluation remain to be defined, and that is not the central question of this book. But it is interesting to now think of an aesthetics of existence as mediated subjectivity with which we ended the last chapter, in terms of reterritorialized nature. The idea of nature as something that is both given and given to control is analogous to the idea of subjectivity as something that is in part constituted by technological mediations and can also act on those mediations. One can say that in the current re-naturalized or reterritorialized understanding of nature as given and given to control, how we are constituted by our biology, by our nature, can potentially be guided, and in desirable ways; that a certain relationship can be stylized in relation to that nature which defines us. ARTs in this context, are technologies of the self, as the examples taken from the ethnographic work shows here. They help create new identities as they are woven into new narratives of nature, parenthood and family.

References

ASRM. (2009). Access to fertility treatment by gays, lesbians, and unmarried persons. *Fertility and Sterility, 92*(4), 1190–1193.

Balsamo, A. (1996). *Technologies of the gendered body: Reading cyborg women*. Durham/London: Duke University Press.

Becker, G., Butler, A., & Nachtigall, R. D. (2005). Resemblance talk: A challenge for parents whose children were conceived with donor gametes in the US. *Social Science & Medicine, 61*(6), 1300–1309.

Birenbaum-Carmeli, D., & Dirnfeld, M. (2008). In vitro fertilization policy in Israel and women's perspectives: The more the better? *Reproductive Health Matters, 16*(31), 182–191.

Blackburn-Starza, A. (2008). Lesbian couple fights NHS over fertility treatment. *BioNews*, (518). http://www.bionews.org.uk/page_45634.asp. Accessed 14 June 2013.

Braidotti, R. (1994). *Nomadic subjects: Embodiment and sexual difference in contemporary feminist theory*. New York: Columbia University Press.

Braidotti, R. (2002). *Metamorphoses: Towards a materialist theory of becoming*. Cambridge: Polity Press.

Braidotti, R. (2006). *Transpositions: On nomadic ethics*. Cambridge: Polity Press.

Brewaeys, A., Ponjaert, I., Van Hall, E. V., & Golombok, S. (1997). Donor insemination: Child development and family functioning in lesbian mother families. *Human Reproduction, 12*, 1349–1359.

Bryld, M. (2001). The infertility clinic and the birth of the lesbian. *European Journal of Women's Studies, 8*(3), 299–312.

Deleuze, G. (1995). *Negotiations 1972–1990* (trans: Joughin, M.). New York: Columbia University Press. Original edition, 1990.

Deleuze, G., & Guattari, F. (1977). *Anti-Oedipus: Capitalism and schizophrenia* (trans: Seem, M., Lane, H.R., & Hurley, R.). New York: Viking Press. Original edition, 1972.

Deleuze, G., & Guattari, F. (1987). *A Thousand plateaus: Capitalism and schizophrenia* (trans Massumi, B.). Minneapolis: University of Minnesota Press. Original edition, 1980.

Deleuze, G., & Guattari, F. (1994). *What is philosophy?* (trans: Tomlinson, H., & Burchell). New York: Columbia University Press.

Deleuze, G., & Parnet, C. (1987). *Dialogues* (trans: Tomlinson, H., & Habberjam, B.). London: Athlone Press.

Derrida, J. (1976). *Of Grammatology* (trans: Spivak, G. C.). Baltimore: Johns Hopkins University Press. Original edition, 1967.

Derrida, J. (1981). *Dissemination* (trans: Johnson, B.). London: Athlone Press. Original edition, 1972.

Didur, J. (2003). Re-embodying technoscientific fantasies: Posthumanism, genetically modified foods, and the colonization of life. *Cultural Critique, 53*, 98–115.

Dolan, M. (2008 Aug 19). California doctors can't refuse treatment to gays on religious grounds, court rules. Los Angeles: *Los Angeles Times*.

Duden, B. (1993). *Disembodying women: Perspectives on pregnancy and the unborn*. Cambridge, MA: Harvard University Press.

Eskridge, W. N., & Stein, E. (1998). Queer clones. In M. Nussbaum & C. Sunstein (Eds.), *Clones and clones: Facts and fantasies about human cloning* (pp. 95–113). New York: W.W. Norton & Co.

Firestone, S. (1970). *The dialectic of sex: The case for feminist revolution*. New York: Bantam Books.

Foucault, M. (1973). *Madness and civilization: A history of insanity in the age of reason* (trans: Howard, R.). New York: Vintage Books. Original edition, 1961.

Foucault, M. (1979). *The history of sexuality, Volume 1: An introduction* (trans: Hurley, R.). London: Allen Lane. Original edition, 1976.

Franklin, S., & Ragone, H. (Eds.). (1998). *Reproducing reproduction: Kinship, power, and technological innovation*. Philadelphia: University of Pennsylvania Press.

Franklin, S., & Roberts, C. (2006). *Born and made: An ethnography of preimplantation genetic diagnosis*. Princeton: Princeton University Press.

Franklin, S., Lury, C., & Stacey, J. (2000). *Global nature, global culture*. London: Sage.

Griggers, C. (1997). *Becoming-woman*. Minneapolis/London: University of Minnesota Press.

Guattari, F. (1992). Regimes, pathways, subjects. In J. Crary & S. Kwinter (Eds.), *Incorporations* (pp. 16–35). New York: Zone.

Haelyon, H. (2006). "Longing for a Child": Perceptions of motherhood among Israeli-Jewish women undergoing in vitro fertilization (IVF) treatments. *Nashim: A Journal of Jewish Women's Studies and Gender Issues, 12*(Fall), 177–203.

Halberstam, J., & Livingstone, I. (Eds.). (1995). *Posthuman bodies*. Bloomington: Indiana University Press.

Haraway, D. (1997). *Modest_Witness@Second_Millenium. FemaleMan©_Meets_Oncomouse™: Feminism and technoscience*. New York: Routledge.

Hardt, M., & Negri, A. (2000). *Empire*. Cambridge, MA: Harvard University Press.

Hartouni, V. (1991). Containing women: Reproductive discourse in the 1980s. In A. Ross & C. Penley (Eds.), *Technoculture* (pp. 27–56). Minneapolis: University of Minessota Press.

Hartouni, V. (1997). *Cultural conceptions: On reproductive technologies and the remaking of life*. Minneapolis: University of Minnesota Press.

Hashiloni-Dolev, Y. (2007). *A life (un)worthy of living: Reproductive genetics in Israel and Germany*. Secaucus: Springer.

Hayles, N. K. (1999). *How we became posthuman: Virtual bodies in cybernetics, literature and informatics*. Chicago: University of Chicago Press.

Heffernan, T. (2003). Bovine anxieties, virgin births, and the secret of life. *Cultural Critique, 53*, 116–133.

Inhorn, M. C., & Birenbaum-Carmeli, D. (2008). Assisted reproductive technologies and culture change. *Annual Review of Anthropology, 37*, 177–196.

Jacobus, M., Keller, E. F., & Shuttleworth, S. (Eds.). (1990). *Body/politics: Women and the discourses of science*. London: Routledge.

Jameson, F. (1984). Postmodernism, or the cultural logic of late capitalism. *New Left Review, 146*(1), 53–92.

Kahn, S. M. (2000). *Reproducing Jews: A cultural account of assisted conception in Israel*. Durham: Duke University Press.

Kahn, S. M. (2002). Rabbis and reproduction: The uses of new reproductive technologies among ultraorthodox Jews in Israel. In M. C. Inhorn & F. Van Balen (Eds.), *Infertility around the globe: New thinking on childlessness, gender, and reproductive technologies*. Berkeley: University of California Press.

Levidow, L. (1996). Simulating mother nature, industrializing agriculture. In G. Roberston, M. Mash, L. Tickner, B. Curtis, & T. Putnam (Eds.), *FutureNatural: Nature, science, culture* (pp. 55–71). London/New York: Routledge.

Liu, C. (2009). Restricting access to infertility services: What is jutsified limitation on reproductive freedom? *Minnesota Journal of Law Science and Technology, 10*(1), 291–324.

Lycett, E., Daniels, K., Curson, R., & Golombok, S. (2004). Offspring created as a result of donor insemination: A study of family relationships, child adjustment, and disclosure. *Fertility and Sterility, 82*, 172–179.

Macrae, F. (2008). Death of the father: British scientists discover how to turn women's bone marrow into sperm. *Daily Mail*. http://www.dailymail.co.uk/sciencetech/article-511391/Death-father-British-scientists-discover-turn-womens-bone-marrow-sperm.html. Accessed 11 June 2013.

Mamo, L. (2007). *Queering reproduction: Achieving pregnancy in the age of technoscience*. Durham: Duke University Press.

Murray, C. (2004). Same-sex families: Outcomes for children and parents. *Family Law, 34*, 136–139.

Newman, K. (1996). *Fetal positions: Individualism, science, visuality*. Stanford: Stanford University Press.

Parisi, L. (2004). *Abstract sex: Philosophy, bio-technology, and the mutation of desire*. London: Continuum Press.

Pattinson, S. D. (2003). Current legislation in Europe. In J. Gunning & H. Szoke (Eds.), *The regulation of assisted reproductive technology* (pp. 7–20). Hampshire: Ashgate Press.

Petchesky, R. P. (1987). Fetal images: The power of visual culture in the politics of reproduction. *Feminist Studies, 13*(2), 263–292.

Ragone, H. (1994). *Surrogate motherhood: Conception in the heart*. Boulder: Westview Press.

Shalev, C., & Goolding, S. (2006). The uses and misuses of in vitro fertilization in Israel: Some sociological and ethical considerations. *Nashim: A Journal of Jewish Women's Studies and Gender Issues, 12*(Fall), 151–175.

Shildrick, M. (1997). *Leaky bodies and boundaries: Feminism, postmodernism and (bio)ethics*. London: Routledge.

Simon, B. (2003). Toward a critique of posthuman futures. *Cultural Critique, 53*(Winter), 1–9.

Sourbut, E. (1996). Gynogenesis: A lesbian appropriation of reproductive technologies. In N. Lykke & R. Braidotti (Eds.), *Between monsters, goddesses, and cyborgs* (pp. 227–241). London: Zed Books.

Spar, D. L. (2006). *The baby business: How money, science, and politics drive the commerce of conception*. Boston: Harvard Business School Press.

Stabile, C. (1994). *Feminism and the technological fix*. Manchester: Manchester University Press.

Storrow, R. F. (2006). Marginalizing adoption through the regulation of assisted reproduction. *Capital University Law Review, 35*, 479–516.

Strathern, M. (1992). *Reproducing the future: Essays on anthropology, kinship and the new reproductive technologies*. Manchester: Manchester University Press.

Strathern, M. (2005). *Kinship, law and the unexpected: Relatives are always a surprise*. Cambridge: Cambridge University Press.

Szkupinski Quiroga, S. (2007). Blood is thicker than water: Policing donor insemination and the reproduction of whiteness. *Hypatia, 22*(2), 143–161.

Thompson, C. (2005). *Making parents: The ontological choreography of reproductive technologies*. Cambridge, MA: MIT Press.

Walker, K. (2003). Should there be limits on who may access assisted reproductive services? A legal perspective. In J. Gunning & H. Szoke (Eds.), *The regulation of assisted reproductive technologies* (pp. 123–140). Hampshire: Ashgate Press.

Chapter 8
New Modes of Ethical Selfhood: Geneticization and Genetically Responsible Subjectivity

Abstract In this final chapter, a mediated posthumanist perspective that incorporates each of the important aspects of various approaches that have been discussed – the non-humanist basis of radical and methodological posthumanism, the schizoanlaytic framework developed by Deleuze and Guattari, and the Foucauldian approach to ethical subject-constitution – is used in an examination of new genetic technologies.

First the "geneticization" thesis is discussed and critiqued within a schizoanlaytic reading. Here the paranoid tendency of genetic determinism and essentialism is always accompanied or contested by schizophrenic narratives of genomic complexity and a novel ontology of flatness. As with the example of ARTs and the category of nature, new categories of "life" and "selfhood" emerge, and we are once more confronted with the paradox of emerging biotechnologies: that biology is both given and given to control.

Secondly, we also encounter the emergence of a new mode of subjectivity – genetically responsible selfhood, which implies that individuals are increasingly defining themselves in terms of genetics but that this "geneticization" is not deterministic, since it is often seen as a resource that can be used to shape one's life in accordance with personal hopes and values. It is argued that this mode of subjectivity is best understood in light of the notion of technological mediation and Foucault's work on ethical subject constitution, or the understanding that subjects can actively relate to and help shape the mediations that constitute them as subjects. More than any other posthumanist approach, mediated posthumanism succeeds best in capturing this new mode of selfhood.

Keywords Geneticization • Genetic responsibility • Technologies of the self • Technological mediation • Ethics

In this final chapter, a mediated posthumanist perspective that incorporates each of the important aspects of various approaches that have been discussed – the non-humanist basis of radical and methodological posthumanism, the schizoanlaytic framework developed by Deleuze and Guattari, and the Foucauldian approach to ethical subject-constitution – will be used in an examination of new genetic technologies, from genome mapping, to neurogenetics and personal genomics. In a first part I discuss what is known as the "geneticization" (Lippman 1992; Nelkin and Lindee 1995), the process by which essential truths about biology, the body, the brain, but also about less tangible notions like personality, behavior and selfhood, are seen to be increasingly articulated through genetic forms of reasoning and treatment. Geneticization arises as an effect of narratives of genetic reductionism and determinism within scientific discourse which are often extended from methodological or epistemic models that look to genes and DNA as a preferred explanatory level, to more metaphysical models in narratives of genetic essentialism in popular discourse. As we shall see however, genetic determinism and essentialism, what appear to be paranoid tendencies in the schizoanalytic terminology, are always accompanied or contested by schizophrenic narratives of genetic complexity and a novel ontology of flatness (rather than of depth), according to which there is nothing mystical or unrepresentable about the "secret of life", and no deep or authentic truth behind selfhood. As with the example of assisted reproduction and its implications for the notion of nature, both of these tendencies and types of processes coexist, giving rise to novel understandings of life and selfhood. We once more encounter the underlying paradox and novelty that emerging biotechnologies point to: that biology is both given and given to control. In the context of genetics and the increasingly biological accounts of identity today, this is articulated in the notion that biology may define the limits of human being, but that it is simultaneously open to virtually *un*limited possibilities, that biology opens up a space of uncertainty.

These interactions, this inherent tension in the new genetics, also expresses a shift in the kind of persons we take ourselves to be, and is giving rise to a new mode of subjectivity: the genetically responsible subject, which will be the focus of the second part of this chapter. Genetically responsible subjectivity implies that individuals are increasingly defining themselves in terms of genetics, as a process of geneticization, but that this redefinition does not lead to a new or reinforced passivity in the face of a genetic fate, rather, it is increasingly seen as subjective potential, as a resource that can be used to shape one's life in accordance with personal hopes and values. Accompanied by other important contemporary transformations including the introduction of the language of genetic risk and the increasing individualization and privatization of social risk in the political realm, genetic responsibility refers to a responsibility towards oneself, via a prudent management of genetic risks (Rose 2007; Lemke 2011).

This mode of subjectivity can be understood in light of the notion of technological mediation, the idea that subjects are constantly being constituted and transformed by their engagements with technology, which I have been pitting against the view assumed by dystopic and liberal posthumanism, that technologies are in some essential way separate from humans. The genetically responsible mode of subjectivity, as we shall see, is not without its critiques, namely, that genetically responsible subjects are disciplined subjects who have internalized hegemonic norms of healthy behavior, or consumers of health services who have become the implementers of the dictates of the market. But this critique does not take into account the profound implications of the notion of technological mediation for personal autonomy and freedom, and draws it back on to familiar categories of freedom and sovereignty anchored in liberal humanism. Furthermore, by emphasizing the disciplinary over and above the creative aspect of subject constitution in Foucault's work, this critique does not identify how numerous engagements with genetic technologies actually resist disciplinary narratives in unexpected ways. Alternatively, this new mode of subjectivity may indicate the emergence of a new ethics of technology that is not based in liberal humanism.

8.1 The Geneticization of Life: From Reductionism to Complexity, and from Depth to Surface (and Back)

8.1.1 From Genetic Determinism to Genetic Essentialism

1953 was somewhat of a miraculous year for biology. It was in this year, in a series of papers, that the structure of DNA was defined (Franklin and Gosling 1953; Watson and Crick 1953a, b). The gene, and more recently DNA as genetic material, had always been understood to be the self-replicating, so-called "stuff of life". But the mechanism by which this replication was secured was somewhat of a mystery. The discovery of the double helical structure, with its base pairing copying mechanism, resolved this, establishing DNA as the molecule that not only holds the secrets of life but also executes its instructions. This double function was articulated in what came to be known as the central dogma of molecular biology, by which a single gene codes for the production of a single corresponding protein molecule, which in turn determines properties of the organism. In the words of Francis Crick (1958), "DNA makes RNA, RNA makes protein, and proteins make us". Though highly disputed since its introduction, this dogma has acted as a guiding image through what Evelyn Fox Keller (2000) has called the "century of the gene".

The wonderful simplicity of the central dogma implies two types of inferences: a deterministic one – that genes determine the properties of the organism, and a reductionist one – that only genes, as opposed to more complex and multiple interactions, shape those properties. From here, deterministic and reductionist frameworks shape several explanatory levels. On the level of the organism determinism is conferred to DNA as the "master molecule". On the level of populations and species, genes are portrayed as the real targets of natural selection and the main actors shaping phenotypic outcomes. The flow of genetic information in the central dogma is unidirectional, so that there is no way for information to travel back into the genetic material, a fact that effectively rules out any environmental effects on heredity. This linearity corresponds to the view of the vertical transmission of traits in evolutionary biology that we have seen in Chap. 5. Indeed, Richard Dawkins' (1989) concept of the "selfish gene", by which DNA is the fundamental unit of inheritance and reproduction – the "replicator" – and organisms are simply "vehicles" for the successful transmission of replicators, has played a significant role in popularizing the reductionist tendency of much of modern biology. In the schizoanlaytic terms developed in the previous chapter, this over-valorization of the gene is an evidently paranoid process, in which all the workings of biological life and all phenotypic properties of an organism can be reduced to one master molecule.

But the very prominence of genes in biological discourse, and the analytic techniques that molecular biology has gradually developed to explore the relationship between genetic structure and biological function (genomics), have rather revealed how vastly complex the secret of life – and how large the gap between genetic information and biological meaning – really are. From its beginnings, developments in the field of molecular genetics have contributed to a deterritorialization of a deterministic view of life. Early on, minor blemishes began to appear on the seamless image of the central dogma, with the discovery of "structural" and "regulator genes" "split genes" and "junk DNA". It soon became clear that one gene can also code for a number of proteins and that one protein can determine a number of functions. In 2001, the central dogma was overtly shaken with the publishing of the draft mapping of the human genome (Venter 2001). In the mapping, only about 30,000–40,000 of an expected 100,000–300,000 genes were found, thus challenging the gene's function as a unique coder for single characteristic traits.[1] In other words, much too few genes were discovered to explain whole-organism traits in simple genetic deterministic terms. It became clear that the concept of the gene really denotes two different entities, a structural one and a functional one, that only emerges out of the dynamic interaction between many different factors.

In the framework of genomics, DNA is both an inert and a relational molecule, that does not "do" anything in itself, but requires other genes and the environment in order to act. Here genes are rather thought of as parts of an ecosystem, where any change in the environment or any mutation or replacement of a gene has the potential to influence every other gene in the system. The vast majority of human diseases, indeed, are multifactorial (diseases like cystic fibrosis, thalassemia, Huntington's chorea and Tay-Sachs, which we hear about most in relation to genetics, are *mono*genic diseases and much rarer). Multifactorial diseases are influenced by a number of interacting genes and a vast array of signals within the cellular environment that include nutrient supply, hormones and signals from other cells, all of which are in turn influenced by the organism's external environment as a whole. Even the metabolic disorder PKU, which is considered to be a single-gene disorder, reflects this complexity: those born with PKU, which can lead to mental retardation and early death, cannot digest the amino acid phenylalanine that is normally present in food. But if they receive a modified diet in which none of this amino acid is present, the defective gene never becomes manifest. Strictly speaking, then, PKU can have a purely environmental cause and need not be genetically determined at all.

This understanding of gene dynamics has led some theorists to speak of a "post-genomic" biology (Bains 2001; Stotz 2006). Here complexity seems to override reductionism, and the focus is less on the gene and increasingly on interactions,

---

[1] By 2004, this number was again reduced to 20,000–25,000 (Collins 2004).

developmental sequences and multi-directional regulation in the synthesis of proteins, such as transcription processes (transcriptomics), variations smaller than the gene such as Single Nucleotide Polymorphisms (SNPs), and the cell and the process of the creation of proteins (proteomics). This echoes some of the transformations discussed in Chap. 5, in the context of evolutionary biology, where developmental systems theory, epigenetics and non-vertical means of gene transfer point to a similar shift. Post-genomic models in molecular biology and what we might call their "post-neo-Darwinian" counterparts in evolutionary biology challenge the view of genes and organisms as closed entities that can be isolated and distinctively defined. They put a much greater emphasis on interconnectedness, interdependence, relationality and openness, and as such offer schizophrenic responses to paranoid narratives of genetic reductionism and determinism. They deterritorialize the central dogma's narrow, unidirectional view of life in which instructions are read out and followed, freeing life, as it were, from this linear, two-dimensional configuration, while weakening the notions of control, predictability and causality that underlie genetic determinism.

Even so, it is difficult to say that the recent thrust of post-genomic research has completely undermined the explicitly deterministic and reductionist assumptions earlier embodied by the central dogma. The privileged role given to genes in understanding "life", the centrality of DNA, as well as linear, causal narratives, remain central to genetic research today despite an acknowledgement of complexity. Brian Wynne (2005) has examined how current genomics science expresses contradictory tendencies of complexity and predictive determinism and reductionism at the same time. While complexity, as the limits of predictability, is continuously encountered, he argues, it is also just as soon "bracketed" or "denied" in favor of expectations of control. Wynne analyzes the continued prevalent use of verbs like "control", "program" or "determine" in scientific discourse, that work to preserve the image of genes as the principle agents in what is nonetheless claimed to be an interaction between genes and environment. Gene-determinism seems to be rescued in this sense by shifting to the idea that the whole genome is the determining single variable, so that special sequences of genes can explain the variety of phenotypical outcomes even if individual genes alone cannot. In this reterritorialization, the loss of control or predictability introduced by the notion of complexity is recuperated to a certain extent, and in this context, determinism and reductionism can coexist alongside complexity models. Here too, just as critics lament the cooptation of hybridity and difference by capitalist rhetoric, it seems wise to differentiate between the vocabulary of complexity as understood by critical theorists and by scientists, where it "remains indebted to the modern project of science" and the search for a new kind of causality (Dillon 2000: 7). As Wynne writes:

> Some of those attracted to "complexity-thinking" appeared to understand it as a new ontology involving moral recognition of the falsehood of ambitions and pretences of prediction and control, thus perhaps as a non-hegemonist epistemology ... [I]n conventional discourse-practices control (with tacit externalization) remains a persistent expectation, as a moral and intellectual given. According to this latter ontology, complexity is simply a complex *object* per se, but one ultimately amenable to control. (2005: 71)

Of course, it is important to distinguish between what we might call a methodological or epistemic determinism and a philosophical or ontological determinism here. One would be hard pressed to find a biologist today who would claim that phenotypical traits are not the result of an interplay of many genes and the environment. The reductionist assumptions of research do not necessarily require geneticists to advocate a universal philosophical determinism that implies that *everything* is genetically or biologically determined. Genetic determinism and reductionism is thus often presented as an insiders' vocabulary, a jargon adopted by scientists as a result of the capacity they have to isolate aspects of genes, that is to say, the use of a specific hierarchical model that they choose for the sake of organizing research, not in order to say anything else about the world.

But is this distinction more than just theoretical? The concept of the gene has always been a complex and dynamic rhetorical creation, cultivated by the language skills of the scientists who have attempted to conceptualize it (Kay 2000; Keller 1995, 2000; Lewontin 1991). The term itself, as the concise review above suggests, is highly varied and imprecise, and can refer to, among others, a coding function, a location on a chromosome, a nucleic acid sequence, or a unit of heredity, just as the term genome can mean all the genes in a single individual, all the genes in a particular species, or simply the draft sequence of a majority of genes within a group of people. This elusiveness has allowed for the age-old belief that there is a "language" or a "code book of life" as well as other metaphorical language (e.g. the "selfishness" of genes), to slip back in to what is held to be a strictly scientific discourse, and has allowed geneticists to enjoy the benefit of metaphysical explanations without paying the price of a loss of scientific credibility. This metaphorical language, as critics commonly argue, is then carried (back) outside of the laboratory into popular discourse, giving the gene its iconic status, as an "obligatory passage point" (Latour 1987: 245) through which current accounts of life, cognition, behavior and identity must pass.

8.1.2   *The Sublime Object of Biology*

This saturation of popular discourse with metaphors from genetics, what Keller (2000) has called "gene talk", has become a main focus of study for scholars from the humanities and the social sciences, who have in recent decades generated an abundance of research on the discourse of geneticists and on the cultural representation of genetics. This has led to a broad critique of the phenomenon that has been coined "geneticization" or "genetic essentialism". Abby Lippman first defined geneticization as:

> the ongoing process by which priority is given to differences between individuals based on their DNA codes, with most disorders, behaviors and physiological variations defined, at least in part, as genetic in origin. … Through this process, human biology is incorrectly equated with human genetics, implying that the latter acts alone to make us each the organism she or he is. (1993: 178)

The term "genetic essentialism" was first used by Sarah Franklin to depict any "scientific discourse ... with the potential to establish social categories based on an essential truth about the body" (1993: 34). In their influential work on the popular accounts of genetic practices in the US, *The DNA Mystique* (1995), Dorothy Nelkin and Susan Lindee argued that the gene has received the legitimacy of an almost supernatural entity in American popular culture, that can define identity, determine human behavior and explain social problems. Donna Haraway has suggested the term "genetic fetishism" to describe the process by which human and non-human liveliness is translated into a genetic vocabulary or "map", and then reified, fetishized, and mistaken for life itself (1997: 143). The gene is fetishized, she explains, when it comes to be seen itself as the source of value, when its abstraction is taken for the concrete entities that it represents. This takes us back to the problematic use of metaphorical language in scientific discourse and how it is carried over into popular discourse, and ties in to Nelkin and Lindee's thesis that the gene has been endowed with the qualities of a sacred object, having taken on the secular equivalent of the soul in our society. Haraway too, remarks that genetic discourse is characterized by a "barely secularized Christian realism" (1997: 10).

The process by which genetic discourse, both inside and outside the laboratory, has (re)localized the essence of our humanity within our DNA, can be viewed as a reterritorialization of the essence of life within a distinctly modernist account of biological life, in which life is construed as a mysterious, invisible force, common to all beings but hidden away in the unseen depths of the body. Foucault traces the emergence of this narrative of life in *The Order of Things* (1989). In the classical age, according to Foucault's archaeology, natural history was interested in establishing a taxonomical comparison of all living beings according to observable characteristics, a conceptual framework in which "life" itself did not exist. But at the end of the eighteenth century, with the introduction of a division of nature between the organic and the inorganic, a new depth opened up beneath the taxonomic table of beings, and the notion of life, as an underlying, invisible unity or vitalism shared by all organisms, allowed for the development of the discipline of biology. In this modern comparison of living beings,

> The differences proliferate on the surface, but deep down they fade, merge, and mingle, as they approach the great, mysterious invisible focal unity, from which the multiple seems to derive ... Life is no longer that which can be distinguished in a more or less certain fashion from the mechanical; it is that in which all the possible distinctions between living beings have their basis. (1989: 269)

Biology, as the science of life, thus came to be characterized by an ontology of *depth* in this account, in which the observable or external traits of organisms, their visible surface, were understood in terms of the set of basic laws that determined them, a vitality that was plunged into the invisible depths of organic bodies. Foucault's argument in *The Order of Things* is that this depth ontology informs the modern episteme, and can be located in various disciplines, from economic thought, where it is embodied by the invisible hand of the free market, to psychoanalysis, where it takes on the form of the psychic interiority of the subject. Molecular biology and contemporary genomics also work within such an ontology of depth. The rearticulation of DNA as genetic code, and of the genome as the "language of life", has been accompanied by a conceptualization of the genome as a deep inner truth, the essence or secret of life that is expressed in the surface of corporeality and behavior. DNA is endowed with quasi-mystical and sacred qualities, that invokes religious and cryptic imagery, from the "Holy Grail" of genetics, to the "Book of Life", to the "Rosetta Stone" and the "lingua franca" of the human organism.[2] Here surface phenomena, or phenotypes, are made intelligible in terms of an underlying interior, the genotype, and to a great extent, "life" in the form of the double helix, remains magical, quasi-immortal (Dawkins 1999), safely embedded in essentialized data.

Alongside this reterritorialization of life's essence as hidden interior, yet another movement of deterritorialization undermines this depth ontology: by claiming to discover the genetic basis of life, genomic research also simultaneously exposes or lays bare what it defines as the essence of life. As Richard Doyle has skillfully argued in his study of the rhetorical transformations of the life sciences *On Beyond Living* (1997), the rhetorical effect of the geneticization of life is a narrative of *resolution* – that the search for the essence of life is finally over. In the telling words of Nobel Prize-winning biologist François Jacob (1973):

> Biology has demonstrated that there is no metaphysical entity hidden behind the word "life". The power of assembling, of producing increasingly complex structures, even of reproducing, belongs to the elements that constitute matter. (306)

If the modernist narrative of life perceived life as a mysterious force that needed to be teased out from the depths of the organism, contemporary biology does precisely that – ousting the "secret" in the form of the genes. "If we used to think our fate was in the stars", James Watson feels confident to claim in 1989, "now we know, in large measure, our fate is in our genes" (Quoted in Jaroff 1989). With the discovery of the structure of DNA and the ongoing sequencing of complete genomes, the question of the secret of life seems to have been resolved in the revelation that there is no secret, that there is nothing invisible, ungraspable or unrepresentable in the phenomenon of life, no hidden essence beyond the codes and sequences of nucleic acids. Doyle views this as a strange new rhetoric of the sublime, "the remains of the sublime, sublime remains whose fascination is tied to the memory of a story that looked for something beyond fragmented surfaces". The sublime object of biology, he writes, "is no longer the life that is beyond disease and the organism, visibly invisible; instead, it is the continual story that there is nothing more to say, a story of resolution told in higher and higher resolution" (1997: 20).

This second rhetorical effect of contemporary genomics, the idea that there is no secret behind the notion of life, works to abolish the distinction between depth and

---

[2] For a detailed account of the use of metaphors of information that made possible the prevalent notion that DNA is the "code" of life, see Lily E. Kay's *Who Wrote the Book of Life* (2000). Kay analyzes the production of the "information discourse" in molecular biology, and the discursive shift which led to the equation of organisms and molecules with information storage and retrieval systems, and heredity with information transfer. See also Susan Oyama's *The Ontogeny of Information* (2000b) for a critique of this process.

surface that the modernist ontology of depth is based on. It establishes that all interior spaces are equally superficial, that depth is only latent surface. This movement from depth to surface can be related as a schizophrenic deterritorialization of meaning par excellence, as one of the recurrent themes of postmodern theory's emphasis on surfaces, connections, networks, superficiality and signifiers as opposed to meaning, value, content, the signified, etc.[3] But as suggested here, there is a constant fluctuation between both these narratives, depth and the underlying genotype as the seat of the essence of life, surface and the revelation that the genetic code, which can be known, controlled and manipulated, is all there is. Similar to the processes of de- and re-naturalization of nature discussed in the previous chapter, these narratives of depth and surface constantly intertwine rather than cancel each other out, and as we shall see further on at other sites where interiority seems to be being flattened out, they often coexist.

8.1.3 *Behavioral Genetics or Genetic Behavior*

If genetic reductionism starts out as a methodological or epistemic model that looks to genes and DNA as a preferred explanatory level, the greater repercussions of this strictly scientific paradigm both within scientific discourse and beyond it cultivates a paranoid trend towards a more general genetic determinism and essentialism that implies several important notions: the idea that the biological meaning of human being is equated with genetic constitution; the idea that the historical and moral complexity of human beings can be reduced to their genes; and the idea that many socially significant traits are largely under genetic control and invariant across a range of environmental conditions of development. It is this form of genetic essentialism that accounts for the overwhelming number of "gene-for" type reports that flood the media, whether regarding genetic diseases or the genetic basis of human behavior and traits, ranging from violence and alcoholism, to intelligence, depression and risk taking.

Perhaps even more than molecular biology, genetic essentialism is fostered by research in behavioral genetics which goes beyond phenotypic traits and attempts to identify which patterns of human behavior are determined by our genes by estimating the percentage of heritability in them.[4] Behavioral genetics assumes that, even if genes cannot directly regulate behavior, they do affect the wiring and workings of the brain, which is the seat of our drives, temperaments and patterns of thought, and

---

[3] For Frederic Jameson (1984) for example, postmodernism signals above all the abandonment of theoretical models of depth and the flattening of spaces into surfaces.

[4] Behavioral genetics is based on the study of twins (and adopted children). Ideally, these studies are carried out on identical twins who have been reared apart, i.e., who share the same genotype but different environments. The assumption is that in such cases, differences in personality or behavior can be attributed to environment. Thomas Bouchard's work at the University of Minnesota is perhaps the most well known example of such research (See Bouchard et al. 1990).

that a substantial fraction of the variation among individuals within a culture can be linked to variation in their genes, whether one is measuring intelligence or personality (Golberg 1993), love (Bartels and Zeki 2000), hate (Zeki and Romaya 2008) or political orientation (Alford 2005). For many behavioral geneticists research has finally demonstrated that one's environment has less of an influence than the unique hand of traits one is dealt in the original genetic lottery at birth. Over a decade ago, the psychologist Eric Turkheimer already ventured to sum up: "the nature-nurture debate is over … All human behavioral traits are heritable" (2000: 160).

Such a claim of course does not take into account the fact that genetic complexity is already inherent to the idea of "genetic influence" and how the notion of genetic complexity undermines the very nature/nurture dichotomy, as mentioned earlier. Furthermore, it is not clear that studies have proven that there is more than "some degree" of genetic causation in higher-order conditions or behaviors. A good example of this is intelligence. If it is believed that a genetic basis for intelligence exists, it also seems that intelligence either involves a number of genes that each have small effects, or that are found in only a small number of people, or both.[5] Another example is happiness. Findings in the rapidly growing field of "happiness studies" (also known as "positive psychology" or the study of comparative "subjective well-being"), indicate that about half of one's subjective happiness is determined at birth by genetics and neurochemistry, and the other half is amenable to positive and negative influence from upbringing, social circumstance, life events, and relationships. This initial brain setting is called the "happiness set-point", and was shortly discussed as a key element in the liberal posthumanist defense of cognitive enhancement therapies in Chap. 3. Similar to this genetic understanding of happiness, in the past couple of decades a number of psychologists have concurred that five key underlying factors that are determined at birth account for most of the variation in personality: openness, conscientiousness, extroversion, agreeableness and neuroticism (Costa and McCrea 1992; Russell and Karol 1994). Studies suggest that the heritability of these factors ranges from 40 % to 60 % (Lykken and Tellegen 1996; Lykken 1999) and here too, while it may not be suggested that behavior is determined by these thermostat settings alone, they are held to influence behavior to the extent that if one's personality is known, good predictions can be made of how that person will respond to a random situation.

8.1.4 The Biochemistry of the Self

Such findings depend on the recent advances in cognitive neuroscience which, along with the field of psychopharmacology, have also extended genetic essentialism, or what we can now call a more general "biological essentialism", further into the spheres of cognition, emotion and mood. This is based on the understanding of

---

[5] The controversy surrounding the genetic basis of intelligence goes far beyond the science of course. See for example the debates concerning the *The Bell Curve* (1994).

psychiatric disorder as being in some way localizable within the body, and more specifically of the recent neuroscientific view that disorders such as depression, schizophrenia, attention deficit hyperactivity disorder, autism and bipolar disorder are localized in the brain.

The notion that some mental illnesses result from a basic neurological or biochemical abnormality which is triggered by some stressful event is not a novel one. Before the reign of Freudian psychoanalysis and its emphasis on biography and experience as the source of mental disorder, the influential neurologist Jean-Martin Charcot was certain that most mental illnesses, including hysteria and epilepsy, were inherited, and only *triggered* by life events (Charcot distinguished between "predisposing" and "occasioning" causes). Much of psychiatry today has returned to this tradition, and supplemented it with clinical observations of the treatment of patients with biological therapies and a host of brain imaging techniques that provide visual proof of the neurological basis and location of disorders. This shift is at the basis of the recent return of brain surgery, which fell into disrepute in the 1950s, for problems like depression, anxiety, some cases of OCD and even obesity, and the use of deep-brain stimulation to treat various affective disorders including major depression.

The most important turning point in this search for the biological underpinning of psychiatric disorders was the momentous discovery of the existence and functioning of a dozen or so chemical neurotransmitters, such as serotonin, dopamine and norepinephrine, that control the firing of nerve synapses and the transmission of signals across the neurons in the brain (Stahl 1996; Greenfield 2000). The levels of these neurotransmitters are believed to affect feelings of well-being, self-esteem, anxiety and the like, and their production and activity is in turn believed to be in part genetic. A "serotonin gene", for example, has been believed to help determine one's risk of depression in response to a stressful life event such as divorce or a lost job.[6] This "neurotransmitter revolution" (Masters 1994) has led to an understanding of mental health in terms of the biological function of the nervous system and of mental disorders in terms of chemical imbalances in the brain.

The theory that emotions are governed by serotonin levels should, theoretically, work just as well the other way around, that is, that emotions and stress levels alter our brain chemistry. But current psychiatric discourse tends to present this relationship in unidirectional terms. In this new way of thinking about mental pathology, the brain and its neurochemistry have become another "obligatory passage point". This is not to say that biographical and environmental effects are completely excluded from research into the etiology of mental illness. As in genomics, a simplified biological determinism is complicated by the view that there is a complex interaction between neurological predispositions and environmental "triggers".

---

[6] The original finding, published in 2003 (Caspi et al. 2003), followed 847 people from birth to age 26 and found that those most likely to sink into depression after a stressful event had a particular variant of a gene involved in the regulation of serotonin. Those in the study with another variant were significantly more resilient. But since then, researchers have been unable to replicate the results on all occasions and in a new study (Risch et al. 2009), the authors reanalyzed the data and found no evidence for the association between a serotonin gene and the risk of depression, no matter what people's life experience was.

Nonetheless, the effects of biographical and environmental factors are relevant mainly insofar as they have an impact on the chemistry and architecture of the brain. In turn, this model shapes research into the production of psychiatric drugs. As Nikolas Rose argues, "the neurochemical brain becomes known in the very same process that creates interventions to manipulate its functioning" (2007: 200), a method of research that has created strong ties between the laboratory, the doctor's office and pharmaceutical companies. Similarly, and already some 20 years ago, Peter Kramer (1993) suggested that mental disorders are increasingly being defined and diagnosed in the clinical setting according to what responds to treatment. In other words, that what a psychiatrist looks for in a patient depends in large part on what medication is available.

Like genetic essentialism, neurological essentialism does not stop at the level of neurological disorders, or faulty neurotransmitter systems, but spills over into the much more blurry realm of personality and selfhood. In the telling words of science journalist Rita Carter, writing about the profundity of the shift in biological psychiatry over a decade ago:

> The biological basis of mental illness is now demonstrable: no one can reasonably watch the frenzied, localized activity in the brain of a person driven by some obsession, or see the dull glow of a depressed brain, and still doubt that these are physical conditions rather than some ineffable sickness of the soul. (cited in Rose 2007: 198)

But she continues in the same breath:

> Similarly, it is now possible to locate and observe the mechanics of rage, violence and misperception, and even to detect the physical signs of complex qualities of mind like kindness, humor, heartlessness, gregariousness, altruism, mother-love and self-awareness.

What might have at one time been identified as the "soul" or the "mind", seem to have been explained away as physical conditions of the brain.

For critics of the biochemical account of mental illness, the attribution of a biological or material basis to qualities that were once deemed immaterial (such as mood, personality and psychological preoccupations) does not only lead to the reduction of a much richer and more complex human identity to their biological infrastructure. It is also contributing to a wholesale "medicalization of personality", a pathologization of variance in moods and traits along a newly drawn axis of normality/abnormality (Conrad 2007; Elliot 2003; Horwitz 2007), as a particularly worrisome aspect of the more general process of medicalization of society (Illich 1975; Szasz 1970). [7] Further, it is suggested that the flip side of this drawing of an increasing number of "personality traits" into the realm of psychiatric health is the creation of new categories in terms of which individuals express perceived problems. Thus the very existence of psychiatric drugs creates new terminological contexts

---

[7] The American Psychiatric Association's Diagnostic Statistical Manual of Mental Disorders is often a target of these critiques. Some of the controversial categories of the current, fifth edition (which contains three times as many disorders and is seven times longer than the first 1952 edition) include "caffeine intoxication disorder", "mathematics disorder" and "sibling relational problem".

within which individuals can increasingly define their troubles and concerns as medical ones. Entire terminologies emerge around specific psychiatric treatments. Ritalin introduces "inattention", "impulsivity" and "restlessness" as categories that are meaningful in identity-formation. And Prozac sheds light on life difficulties as "mild depression" and "anxiety"; allowing people to identify themselves as "catecholamine persons or 5HT persons" (Healy 1997).

8.1.5 Is Selfhood Still Authentic When It Is Technologically Assisted?

Like the narratives of life in genetic discourse, the neuro-biological account of mental states and disorders also implies a shift from an ontology of depth to one of surfaces that might indicate a passage from a modern to a postmodern episteme. As discussed, in Foucault's categorization of epistemic forms, in the modern episteme external or surface phenomena were explained in reference to an underlying depth. The psychiatric gaze and its account of mental illness that emerged in the nineteenth century also located neurosis and psychosis in the deep, inner space of each individual – in what became popularized with Freud as the unconscious, the profound repository of biography and experience. Here too, the body and many of its ills came to be seen as the exterior form or expression of the subject's psychical interior, a hidden truth that no longer lingered on the surface and could be observed by the physician, but had to be spoken, confessed and revealed (Foucault 1979).

In the modern, psychoanalytic account of subjectivity, a psychical interior is understood as the introjection of the body and its parts and the subject's truth becomes what is private, interior and deep in the individual. In psychoanalysis, every symptom has hidden with it, in secret code, its cause, both the historical cause in the early emotional trauma, and the ongoing active cause in the form of the unconscious conflict arising from the mind's attempt to repress that trauma. The route to recovery involves bringing the unconscious struggle into consciousness, i.e., setting the deep hidden truth free. Perhaps the most illustrative symbol of this metaphysics of the interior is the transformation of mental illness from what could be seen to what could be heard and interpreted, a passage from the *eye* to the *ear* as that which can reveal the truth.[8] In this understanding, we are "profound depths, subjects of hidden interiority", while "bodily markings can be read as symptoms, signs, clues to unraveling a psychic set of meanings" (Grosz 1994: 138–139).

As many theorists, including and before Elizabeth Grosz have argued, there has been a close correlation between the psychoanalytic endeavor and the production of the essential, individual, prior self of modern subjectivity, and the use of the metaphor of depth has done much to render this understanding of subjectivity

---

[8] Rose (2007) argues that the visual image analyzed by the trained eye of the physician lost its priority with Kraepelin and Freud, in whose collected works one cannot find a single picture of a patient.

commonsensical. The creation of this psychic depth has thus been taken by many as a key site of critique, and, like its counterpart in the biological sciences, it can be understood as a paranoid attempt to stratify and essentialize, to territorialize, an ill-defined, dynamic vitalism. Indeed, for Deleuze and Guattari (1977, 1987), more than any other modern narrative, it is the psychoanalytic account of consciousness that represses the dynamic investments of desire and channels them into a unifying "molar" whole which fosters the illusion of individuality and produces subjects that conform to the law and authority.

Contemporary psychiatry and the proliferation of neuro sub-disciplines deterritorialize this deep space, expelling or leveling out, so to speak, the interiority of the psyche and offering a narrative of surface connections between neural pathways instead. Here, neurological explanations of disorders are given preference over biographical or subjective ones, and symptoms are detached from their origins, or rather the distinction between symptoms and origins collapses. The popular images of synapse and neurotransmitter mechanisms and an increased use of brain imaging techniques most recently in the form of fMRI and EEG, convey a belief that it is now possible to visualize the interior of the human brain and observe its activity in real time. The "dim glow of the depressed brain", the "frenetic activity of the obsessive compulsive brain" – these visualizations, which in an ontology of depth might have been interpreted as corporeal symptoms of deeper problems related to the unconscious, come to be understood as the disorders themselves. And because in psychoanalytic terms psychic interiority, as the home of the unconscious, was also the seat of identity and subjectivity, the shift to a surface epistemology has the effect of undermining those notions as well, of challenging both their fixity and their transcendental character, and rendering notions of "authentic" and "true" selves highly problematic. What critics of medicalization often bemoan as the loss of the "soul" or of the "spirit" in biological accounts of behavior and personality is precisely this exteriorization or flattening of the modern commitment to a latent interior truth that they argue can never fully be grasped in solely scientific terms.

Alongside this movement of deterritorialization, the model of the brain advanced by neuroscience and psychiatry today and its accompanied belief that phenomena like mood and personality have a biological infrastructure, contributes to a deterritorialization of the notion of nature as something that is given and immutable, insofar as these processes can be intervened upon. As is the case in the context of assisted reproduction, relations between nature, culture, biology and the individual are loosened in this framework. Thus if the neurobiological frame does imply a disturbing paranoid moment that can be seen as reducing or dismissing the richness and complexity of human identity to a biological foundation, this infrastructure, or at least its effects, are no longer deemed to be inevitable, i.e. determinative. Rather, aspects of the individual such as mood and personality are reconfigured as matters of technologically assisted choice or selection (Fraser 2001).

What we might expect with this de-naturalization, as might have been expected of the term "natural" in the context of assisted reproduction, is a fall into disuse of expressions such as "authenticity", "rightful", and "genuine" in relation to personhood and subjectivity. What use do they really retain in this context? But, as with assisted reproduction, these terms are reterritorialized into new understandings of selfhood, where elements of an ontology of depth and a terminology of authenticity coincide with the de-essentialization and flattening out of subjectivity. In *Listening to Prozac* (1993), Peter Kramer's observations of numerous patients who were prescribed anti-depressants are very telling in this sense. As Kramer recounts, patients of his often spoke of "being themselves again" while taking Prozac, or "no longer themselves" off it, and of having located a self that is, "normal," "whole" or "true". Of Tess, one of the patients he writes about most, Kramer explains:

> Tess's reaction to the return of her symptoms when she was off Prozac was: "I'm not myself". But many patients stress a continuity of self on and off drugs. "I am myself without the lead boots", "myself without swimming through Jell-O", "myself on a good day, although I never had days this good", "myself without fears". (195–196)

Other such examples include his patient Sally, who claims that Prozac let her "personality emerge at last", and that she had "not been alive before taking an antidepressant" (148).

Much like the "helping hand" rhetoric surrounding ARTs, what psychopharmaceuticals offer in a framework of neurobiological accounts of well-being is the restoration of a "real" or "normal" self, not the creation of a completely "new" one. These drugs are naturalized, as a sometimes necessary adjunct to the maintenance of an originary, essential self – even as the very fact of their efficacy greatly destabilizes these notions. This lexicon of authentic selfhood also seems to be common among users of neuro-feedback treatment. This increasingly popular technique, used in cases of depression, anxiety, brain damage, attention and learning disorders, among others, uses monitoring devices to display real-time brain activity to patients, who are taught to "train" their brains to remain within a designated range of brain activity. Interviews of users of neuro-feedback reveal similar statements about "real" and "restored selves" (Brenninkmeijer 2010). Patients speak of feeling like their "old self again, only better", and parents speak of the "return" of a beloved child from before the onset of depression, a child who claims to be "myself again" thanks to the use of neuro-feedback.

These narratives seem to incorporate contradictory meanings of subjectivity. On the one hand as that which denotes a deep, unchanging, indefinable but unique individuality, and on the other, as a malleable, tangible entity that can be reduced to neurobiological conditions, is localizable in the brain and can be worked on with technological means. As in the context of ARTs, these narratives do not cancel each other out, but manifest a "digital" quality (Franklin and Roberts 2006) by which they can at different times be emphasized or downplayed. In Brenninkmeijer's study, users seem to effortlessly move between the pronouns "I", "you", "your brain" and "it"; at times creating a distinction between the user and her brain, at times equating these, and at times constructing the brain as more powerful or "willful" than the subject herself, insofar as successful feedback training involves "shutting down" self-consciousness and allowing the brain to condition itself, and insofar as the changes that the brain achieves "on its own" aim to improve "the self". It would be easy to write these off as semantic inconsistency on the part of users. But it would rather be the fixation with semantic consistency on the part of philosophers and social theorists which is unhelpful in these cases.

Here too, empirical, small-scale studies offer important insight and seem to indicate that it is not all clear that the neurosciences have a straightforward influence on accounts of subjectivity (see for example Martin (2010) and Pickersgill et al. (2011)). A tendency to use neurobiological narratives as resources to different ends in different contexts seems to run parallel to the strategizing of users of ARTs. Pickersgill et al. (2011), for example, question the extent of the salience of neurological explanations for individual subjectivity, and suggest that the brain, for many, might be seen as an "object of mundane significance" rather than the locus of identity. Neuroscientific concepts, they found, compete with and are combined with alternative concepts and ontologies (biological, psychological, social, genetic, etc.), that are drawn upon and pieced together by individuals when they theorize about their selfhood. The authors' suggestion that these individuals be seen as "bricoleurs" resonates with the flexible and strategic mobilization of narratives of natural birth and parenthood among users of reproductive technologies.

In this context, then, the paranoid tendency to essentialize subjectivity in a depth ontology is countered and deterritorialized by a schizophrenic tendency to denaturalize subjectivity and stress its inherent transformativity. But these tendencies coexist, and are reterritorialized in a rich, complex account of "neurological selfhood", in which the brain, for many, is "something" but not "everything", and neuroscience an important means of knowing oneself, but not the only one Pickersgill et al. (2011). Like re-naturalized nature, selfhood increasingly incorporates both depth and flatness, essentialism and transformativity. Notions of authentic and real selves persist – even when this self can no longer be the referent for a deep hidden truth in an era where behavior is directly mapped onto the brain – and even when authenticity can no longer be the referent for some pure, originary natural state when nature is constantly being de-naturalized. As in the contexts of ARTs and genomic research, reterritorialization does not imply a return to the original territory, rather, narratives of authentic and real selves can be seen as pockets of depth perforating the surfaceness of neurological explanations of mental states and behavior in a novel, undulating landscape where a variety of accounts of selfhood are layered and overlap.

8.1.6   *Biology Under Control: Biopolitics and Molecularization*

In Chap. 5 I suggested that a "molecularization" of the life sciences – in which life is increasingly understood in terms of genes, proteins and enzymes – denotes a possible shift from a molar to a molecular model of the body, which views the body as an assemblage of discrete and transferable elements rather than a self-contained, unified organic whole. In the context of biomedicine, this process of molecularization rearticulates pathology in terms of sequences of nucleotide bases and specific locations on chromosomes. Diagnostically, this has led to the identification of more and more particular varieties of diseases, or subtypes of diseases, according to their molecular specificities, and in terms of research and treatment, to the attempt to isolate specific molecular structures that can disrupt, restore or activate mechanisms that take place at the level of genes, chromosomes and proteins. Treatments for cancer, for example, are increasingly being targeted at the molecular bases of cancer subtypes following the identification of genetic mutations. The BRCA1 and BRCA2 genes, for example, have been related to an increased risk of certain forms of breast and ovarian cancer. And the inactivation of the von Hippel Lindau tumor suppressor gene, has been related to the stimulation of growth factors which promote angiogenesis in renal cell carcinoma. Likewise, a key aspect of contemporary diagnostic psychiatry is the identification of more and more particular varieties of mental disorder, and the increasingly specific and selective characteristic of its interventions. Here too, research aims mainly at isolating compounds whose specific molecular structure can act on the functioning of a specific neurotransmitter system, by inhibiting specific reuptake pumps or binding onto particular receptors. Thus the new generation of antidepressants that were introduced in the 1990s, such as Prozac, Zoloft, Paxil and more recently Cipralex, are known as "selective serotonin reuptake inhibitors" (SSRIs) which block the reabsorption of serotonin by the nerve synapses in order to increase the levels of serotonin available in the brain: a targeted molecular treatment for a molecular imbalance.[9]

As we have seen, however, molecularization does not only assume that molecular entities and mechanisms of life can be identified and isolated in greater and greater detail and specificity. It also assumes that these entities can be mobilized and manipulated. This is a new kind of involvement in the physiological mechanisms of health and disease, which, with the development of new techniques of intervention such as gene cutting and splicing and polymerase chain reaction, confer a new power or control over the processes of life. At this molecular level, writes Rose, "it seems there is nothing mystical or incomprehensible about our vitality – anything and everything appears, in principle, to be intelligible, and hence open to calculated interventions" (2007: 4). In combination with the process of genetic essentialism, which draws ever more areas of what it means to be human under the scope of biology, this new level of control over biology can be interpreted as the height of modern biopolitics, the attempt to manage and control a population via its biological makeup.[10] In this sense, the current molecularization of biology and biomedicine can be seen as a new means for allowing governments and corporations to establish

---

[9] Indeed, Prozac gained its iconic status as a "smart drug" or a "clean drug", not because it treated depression better than first generation anti-depressants, but because, unlike previous antidepressants which affected three neurotransmitters at once, it concentrates solely on serotonin, thus purportedly preventing a number of side effects.

[10] Foucault proposed the concept of biopolitics in the first volume of *The History of Sexuality* (1979) to designate the expansion, beginning in the eighteenth century, of politics to the management of life in the name of the well-being of the population as a vital order and of each of its living subjects. Biopolitics, he argued, focuses on two poles or realms of intervention: the anatamo-politics of the human body, which seeks to maximize each individuals' labor force and integrate it into efficient systems, and the biopolitics of the population, which focuses on the collective well-being of the general population. Specific problems that became the focus of biopolitics were the size and quality of the population, reproduction, sexuality, health and death.

strategies of regulation, management and knowledge-production over individual and collective bodies, in addition to the techniques of demography and population genetics. And if the ability to intervene upon the vital characteristics of human existence has always been the goal of biopolitics, molecularization has made this possible at an ever more basic and foundational level of life – with the ability to manipulate those molecular entities to boot. As Paul Rabinow speculated in the early days of Human Genome Project,

> The object to be known – the human genome – will be known in such a way that it can be *changed*. This dimension is thoroughly modern; one could even say it instantiates the definition of modern rationality. Representing and intervening, knowledge and power, understanding and reform, are built in, from the start, as simultaneous goals and means. (1992: 236, emphasis in original)

In other words, the administration of life is now not only concerned with populations as it was in nineteenth century biopolitics, but also bound up with the very biology of life itself.

A number of social theorists (Armstrong 2008; Bunton 1997; Nettleton 1997; Petersen 1996) have examined how health discourses enact instances of Foucauldian biopower, and how, as populations have grown and become more fluid and complex, nation states have become "more concerned about the management of life (bio-power) and the governing of populations" (Howson 2004: 125), particularly in relation to health, disease, sexuality, welfare and education. Continuing from Foucault, these theorists argue that if on the surface biopower works in the interest of humanity, on a deeper level it serves other, more nefarious interests, including the control of "deviant" populations. Such a cooptation, so to speak, of the molecular biosciences by large-scale governmental or corporative interests, is not so straightforward, as we shall see shortly. But for critics of genetic essentialism, this has significant repercussions concerning individual citizens insofar as genetic knowledge can be potentially used in discriminatory ways, a kind of genetic discrimination against individuals who have – or worse, who are at *risk of* developing – a genetic disorder, leading to a reduction of social tolerance for human difference. This has been the subject of much debate in relation to insurance and employment (Nelkin and Tancredi 1994; Hubbard and Wald 1997). According to Nelkin and Tancredi, the use of biological tests and genetic screening in the workplace and by insurance companies may possibly create a new "biological underclass" defined as unemployable and uninsurable.[11] Another concern here pertains to projects of preventive screening of youth at risk of violent or aggressive behavior, and the expansion of government strategies of control based on surveillance that would have a genetic basis.

Interestingly, one finds an increasing use of Foucault's term in scientific literature and journalistic texts today in regards to emerging biotechnologies. But in this current use the term is often stripped of much of its historical and critical dimension and refers merely to the social and political implications of biotechnological interventions. See, for example, the liberal posthumanist Institute for Ethics and Emerging Technologies (IEET) website http://ieet.org/, which I have referred to several times, where "biopolitics" has become quite a meaningless buzz word.

[11] In the US, the Genetic Information Nondiscrimination Act was signed into law by President Bush in 2008, but it does not prevent life insurance or disability insurance companies from using genetic information.

For critics of genetic and neurobiological essentialism the medicalization of personality works in parallel to the medicalization of the social realm. This is to say, by privileging studies into genes and neurotransmitters over sociological investigations into the social determinants of psychopathology, what were once considered social problems are being transformed into biological, and individual ones.[12] As mentioned, the reduction of psychiatry to neurobiology tends to neglect phenomenological insights and biographical accounts of the person, or at least to reduce them to the impact they have on the brain. In this model, it is argued, the roots of mental distress are confined to the individual, and the role of social, cultural or political contexts surrounding the person are minimized. Abby Lippman writes, "The individual affixed with a genetic label can be isolated from the context in which s/he became sick ... The individual, not society, is seen to require change; social problems improperly become individual pathologies" (1992: 1472–73). This is especially the case concerning scientific research into the nature and etiology of addiction, violence and crime (Nelkin and Lindee 1995; Rose 2007). Geneticization is thus seen as redirecting much needed resources away from social solutions to social issues, and towards funding of gene mapping projects. For some critics, as we have seen, these concerns run deeper than this. Insofar as medicine always reflects certain conceptions about what normal behavior or personality traits are, and that these conceptions will always be shaped by dominant groups, the shadow of a "new eugenics", or of the use of psychotropic drugs as control strategies against deviant behavior and for the production of compliant subjects (or of a combination of genetic/moral enhancement as per Savulescu (2008)), looms large.[13]

8.2 Genetic Risk and Technological Mediation: The Genetically Responsible Subject

The underlying concern of claims of the geneticization and the molecularization of life is that the richness and complexity of what it means to be human is being diminished to a knowable, controllable biological substratum, according to which life is

---

[12] Camilla Griggers (1997) thus notes that Peter Kramer barely stops to question to what extent Tess's "social style" is a common expression of many women who have experienced abuse in childhood – an important aspect of Tess's biography that he does not care to linger on other than a short sentence in the opening of Tess's case study (Kramer 1993: 1).

[13] Kleinman (1988) and Parens (1998) have argued for example that the success of antidepressant therapies lies in its ability to boost conformity and compliance to advanced capitalist values. This is a point that even Kramer, who is generally seen as supporting the use of Prozac, raises in his book:

> Prozac highlights our culture's preference for certain personality types. Vivacious women's attractiveness to men, the contemporary scorn of fastidiousness, men's discomfort with anhedonia [the loss of the capacity to experience pleasure] in women, the business advantage conferred by mental quickness – all these examples point to a consistent social prejudice. (1993: 192)

no more than DNA and molecular interactions, and individuals are no more than the passive vehicles of a pre-determined genetic fate. It is interesting that this is an argument that is voiced in both dystopic and radical posthumanist discourse. These are pressing concerns, since, as I have argued, the tendencies towards reductionism and determinism *are* characteristic of genetic and neuroscientific research today. But, as I also argued, other elements play into this reterritorialization of human nature as well, in terms of new epistemological frameworks of surfaces in the context of the genetic basis of biological life and the biological basis of personality. Moreover, the genetic make-up of the individual is increasingly viewed as representing subjective potential, not merely objective fate, so that biology, in light of the new level of control that ensues from the molecularization of life, is not so much viewed as an end point, but as a starting point. Nikolas Rose writes:

> In the new field of biopolitics, where interventions are scaled at the molecular level, biology is not destiny but opportunity – to discover the biological basis of an illness, of infertility, of an adverse drug reaction in a cascade of coding sequences, protein syntheses, and enzyme reactions is not to resign oneself to fate but to open oneself to hope. (2007: 51)

Genetic knowledge, as I will argue in the second part of this chapter following Rose and several other theorists, has become a key reference point from which a new mode of subjectivity that implies moral duties and novel fields of responsible action has emerged. This is a new understanding of selfhood that is constituted from within the interactions with emerging biotechnologies that individuals engage in today, and that is best accounted for from a mediated posthumanist perspective.

8.2.1   The Biopolitics of Genetic Risk

Critiques of genetic determinism and essentialism, of "biology under control" and of a new eugenics, do not adequately capture the complexity of the genomic age. Three main distinctions make for this inadequacy: the first one, of a *political* nature, pertains to the fact that biopolitical intervention is no longer a state-sponsored effort aimed at improving the quality of whole populations, like it was in the nineteenth and the early twentieth centuries. The second one, of what can be called an *epistemological* nature, has to do with the introduction of a language of risk as a framework for thinking about health and disease. And the third one, of a more *philosophical* nature, pertains to a shift in the kind of persons we take ourselves to be that is emerging alongside these novel technologies. It is this last shift which is of most interest to any discussion of the notion of the posthuman and the repercussions novel technologies have on new modes of subjectivity. But it must be understood within this larger political and epistemological context, which is described briefly here.

The eugenic projects of the twentieth century that were widespread across Europe and North America and reached their height in Nazi Germany worked within a concern to improve the quality of the human stock in the population as a whole, in the service of a biological struggle between nation states. These operated via a series of coercive policies and practices to eliminate defective genes from the population, which involved compulsory sterilization and the institutionalization of the so-called mentally defective as well as public health measures designed to breed "better" babies (Kevles 1995). As discussed in Chap. 3, for liberal posthumanists it is the coercive and state-sponsored nature of these eugenic programs that make them morally wrong, and the right to reproductive freedom, as the freedom to pursue *individual* conceptions of the good life, that makes "liberal" eugenics morally permissible. It is the neutrality of the state here that ensures this moral legitimacy. Most critics of geneticism, of both the dystopic and radical posthumanist type, admit that a new ("privatized", "free-market" or "consumer") eugenics, would not entail some state-imposed master conspiracy. There is an acknowledgment of the novel roles both individual and commercial interests have in the shaping of this contemporary biopolitics. Indeed, as we shall see shortly, it is the encounter between the "old eugenics" and "the new consumerism" (Sandel 2004) which is of greatest concern here.

However, neither of these views on the new eugenics does enough to account for how significantly different this so-called new eugenics in advanced industrialized societies is, or what I have been calling here molecularized biopower, compared to older forms of biopolitical rationale. On the one hand, as liberal posthumanism upholds, the authoritarian state that would impose its view of fitness on a population is no longer seen as a legitimate form of government. But national health is still a main concern of contemporary states and the object of state-funded preventive medicine and health education today, even if such efforts are no longer framed in terms of the quality or fitness of a population in the field of an international struggle between sovereign states, but in economic terms, such as the cost of treatments, rehabilitation and sick days for the national health system and insurance providers. The liberal posthumanist support of a liberal eugenics does not take into account that the void that has been left by more coercive forms of power has been filled by new forms of self-government that often internalize both the interests for public health of a relatively absent state as well as the commercial aspirations of biotech companies, food retailers and the private health industry. On the other hand, critiques of the new eugenics – while acknowledging the non-coercive means of self-government that contemporary biopolitics employs – still work under the presumption that contemporary biopolitics aims at purging the population of elements that diminish its overall quality, and at legitimating inequality, when the focus here is rather on fostering individual life by maximizing life chances and intervening on inequality by promoting and managing health. Molecularized biopolitics works within a very different space in advanced industrialized societies, where knowledge, power and subjectivity have entered new configurations (Lemke 2002, 2005; Rabinow and Rose 2006; Rose 1998, 2007), that neither of these views succeeds in articulating.

One reason for this is that contemporary biopolitics needs to be understood as *risk* politics. Genomics research today frames illnesses as conditions that arise out of interactions between multiple coding regions, where gene expression can be activated and inactivated by a number of factors and on many levels. Thus, while genes are no doubt prioritized in the discourse of the biosciences, they are framed as vulnerability factors rather than determinants, which render individuals susceptible to certain disorders. Few genes are indicators that a condition will most certainly develop, such as the gene for Huntington's disease. Instead, most genes linked to disease have an effect on the *odds* of developing that illness. Genetic screening, as well, for genes that are directly related to specific diseases like cystic fibrosis and thalassemia that prospective parents commonly undergo today, is also construed in a language of risk: if both parents have a copy of a recessive disease gene, there is one chance out of four that their child will develop the disease. Likewise, types of prenatal screening today are typically statistical tests, which give a statistical estimate as to the chances of the fetus developing a chromosomal defect or malformation. Within the prenatal screening sequencing, fetuses are assigned low-, intermediate- and high-risk scores according to which it is recommended or not to consider further, possibly more invasive testing. While susceptibility, in the form of the correlation of hereditary, age, weight, or dietary factors, has been a basis for medical intervention in the past, the novelty here is that susceptibility can be defined at the level of the individual genome. What genomics, genetic testing and screening establish, then, is not so much a deterministic account of illness, but a *probabilistic* one, that opens up the future to a new kind of calculability. Risk, in this biomedical framework, is a future-oriented mode of thinking that brings a potential future into the range of control of the present.

Consequently, as Rose, Rabinow and Lemke argue, Foucault's concept of bio-power needs to be updated and reinscribed into the political economy of risk management, as we move away from a model of surveillance of individuals and groups known to be a threat to the population, nation or race, toward a model of projecting risk factors.[14] In this "post-disciplinary" society, the identification of individuals or groups for whom genetic risk is seen to be high, coupled with the treatment and management of these risks, replaces the state-sponsored attempts to classify and eliminate or constrain individuals, or conversely, to promote the reproduction of certain individuals in the name of "racial hygiene". Furthermore, in a framework in which all individuals are subjected to the dictate of the genes and all individuals are affected by genetic risk, regardless of socio-economic background, education and working conditions, the eugenic focus on the purification of a collective genetic pool no longer makes much sense, and gives way to an emphasis on the administration of individual genetic risks.

---

[14] Ulrich Beck's *Risk Society* (1992) is usually upheld as the founding work in risk discourse. Beck's thesis is that one can discern a break after the Second World War which marked a new configuration of social groups and their interests not around the organizing principle of industrial production, but around the risks, namely environmental, generated by industrial production. A little before the publication of Beck's book, the French sociologist Robert Castel (1981) also claimed that changes in the concept of risk were indicating the dissolution of modernist society. For Castel trends in the biosciences were particularly relevant to this transformation, notably, a new focus on preventive care and the management of populations at risk and an emphasis on individual work on oneself. Rabinow, Rose and Lemke use this concept of risk to argue for a "post-disciplinary" society.

8.2.2 Genetic Responsibility, Mediation and Moral Decision-Making

At the heart of contemporary biopolitics, then, we find a "genetic governmental-ity" (Lemke 2004), the government of genetic risks, which works via a new individualized genetic responsibility. Again, this transformation does not owe its novelty entirely to the scientific and technological developments in genetics research in recent decades, but must also be understood within a larger political landscape, in relation to the withdrawal of the welfare state and the success of neoliberal programs which have increasingly individualized and privatized the responsibility for social risks. Thus the discourse of genetic responsibility is embedded in a more global discourse of responsibility in Western liberal societ-ies, in which the responsibility for social risks such as illness, unemployment, poverty, etc. has shifted from the public domain of social security to the individual domain of self-care and self-regulation (Petersen and Lupton 1996; Petersen and Bunton 2002; Harris et al. 2010). Geneticization, as we have seen, plays into this shift by recoding social and economic determinants as *biological* risks – one of the main critiques against geneticization – thus turning the source of genetic risk inwards, making it the categorical responsibility of its carrier. For example, mak-ing the case for the socio-cultural root of aggressiveness becomes difficult when there are claims for a genetic basis for violence. This genetic responsibility is increasingly interpreted in terms of responsibility towards oneself, via a prudent management of genetic risks informed by genetic information. But what we also have here, in addition to these political and epistemological transformations, is another more philosophical shift, the emergence of a new mode of subjectivity: the genetically responsible subject. I suggest that this new mode of subjectivity is best articulated by a mediated posthumanist approach.

The genetically responsible subject is the basic unit of reference of contemporary biopolitics; a new category of selfhood that indicates a shift from modern, liberal humanist subjectivity and its two most recent articulations – on the one hand the genetically determined posthuman stripped of her unique human nature forewarned by dystopic posthumanism, and on the other, the autonomous, empowered posthu-man who is master of her destiny professed by liberal posthumanism. Rather, this subject emerges as a technologically mediated, interrelated subject who is guided by principles of self-care and the understanding that one can have an active relation-ship with the technological mediations that help constitute the self.

The notion of technological mediation, as has been discussed, implies that tech-nologies help shape the relation between humans and their world, and that in shaping this relationship, they help to constitute both the objects that are being experienced and the subjects that are experiencing them. As we have seen, in this framework the modernist understanding of subjects as intentional agents and objects as passive matter is undermined since both participants in this relation mutually constitute each other. Technology, or technologies, in this approach, always estab-lish a relation between those "using" them – though "use" here takes on a much less one-sided significance – and their environment, and they enable users to live experiences that may not have been possible before. They thus contribute to the shaping of one's lived experience. For radical posthumanists, the greater implication of such originary prostheticity is that that the subject (or nature) has never been a pure, organic whole, but has always been open to and co-constituted by its environment and tools – and this is a significant step towards a mediated posthumanist understanding of subjectivity. For methodological posthumanists, the greater implication of technological mediation is the redistribution, or delegation of agency and intentionality to and over technical artifacts and what this involves for human beings' interactions with them.

For both these approaches originary prostheticity and technological mediation imply that the view, assumed by dystopic and liberal posthumanism, that technologies are in some essential way separate from humans, or worse, that they are neutral instruments or intermediaries, is flawed. Rather, technologies are active mediators that help shape the relations between humans and reality, and bring about a transformation of the "user" (recall Latour's well-known example of the gun-bearer). For both these approaches, the dystopic posthumanist view that an autonomous, technology-out-of-control no longer has any ends except its own development, and the liberal posthumanist view that technologies in themselves are originally neutral and become what users make of them, are two variations on the same theme. Instead, the understanding that humans are always already technologically mediated implies that the there are no "masters" – neither human nor technical.

Mediation, as we have seen, can take on two forms. A pragmatic form, elaborated by Bruno Latour in his works, that pertains mainly to action, or how specific technologies encourage people to act in certain ways, and a hermeneutic form, exemplified in Don Ihde's works, that pertains to how technologies structure human perceptions and interpretations of reality. Genome sequencing technologies are a particularly illustrative example of both forms of mediation. First of all, the rhetoric of the new genetics assumes that individuals should consult experts who can quantify the risks to their health and act upon this advice. This is clearly a pragmatic form of mediation. Indeed, the discourse of risks embodied by the new medical genetics is indicative of a shift from a relatively reactive or passive position regarding uncertainties like health risks, to an *active* one. The genetically responsible subject is not conceived as someone who passively benefits from medical knowledge and advice, but as someone who actively seeks out information and services. Genomic technologies are also illustrative of hermeneutic mediation, insofar as they help shape the relations between subjects and reality, by contributing to the perception and interpretation of reality. Genome analysis and genetic diagnostics do not provide a "neutral", or "disinterested" representation of one's state-of-health, but generate a reflexive relationship between one's genetic risk profile and social expectations.

Genetic technologies can thus be seen as having a material intentionality, as discussed in Chap. 4: they contribute to the shaping of human interpretations, experiences and decisions. This is to say that the decisions made in light of the knowledge obtained from genetic technologies cannot ever be considered purely "human" decisions – nor, for that matter, can the subject who acts and makes those decisions. On the other hand, this is not to say that so-called subjective decisions and actions are determined by technologies either. It means that such action and decision-making is a *joint* effort of both human beings and technologies.

Subsequent action following the obtainment of genetic knowledge comes in the form of general practices such as genetic testing, periodic screening and monitoring, preventive treatment regimens, and lifestyle changes such as diet and exercise, or more specific and invasive courses of action like the use of PGD, amniocentesis or prophylactic surgery, for example in the case of a heightened risk of breast cancer. In this sense, the decisions and actions taken in the framework of genetic risk concern the realm of *moral* decision-making. This is clear as it becomes obvious that actions and decisions based on genetic knowledge extend beyond the immediate medical and individual realm, as we shall see shortly, and contribute to reshaping prudence and obligation vis-à-vis relations with other family members, marriage, reproduction and broader lifestyle choices. In other words, the mediating role of genetic technologies helps to form the basis of moral decisions taken about the future.

This point is of particular importance for the claim that a new mode of subjectivity is emerging with the use of these technologies. Latour (1992, 1994, 2002), and other methodological posthumanists following him, is known for provocatively arguing that a substantial part of human morality rests upon technological apparatuses (what he calls the "missing masses of morality") and that moral agency is hence distributed over both humans and non-humans. In traditional sociology and philosophy of technology, Latour argues, technologies have always belonged to the realm of means and morality to the realm of ends, and in order to recover our morality and our humanity (lost at some point when technology, or the technological way of being, came to dominate), it is argued that humans must break away from the rule of technology and rediscover a non-technological or non-instrumental way of being in which the sovereignty of ends, not means, can once more reign (a position which is reminiscent of dystopic posthumanism). But for Latour, this view is deeply flawed: the emergence of humanity and the emergence of technologies have been from the outset inseparable, so that there really could be no such thing as a humanity torn away from its technologies, let alone a humanity torn away from its technologies that would be a "moral" as opposed to an "immoral" humanity. According to Latour (2002), once technologies are understood as a particular form of exploring existence and being (among many others), and not only within a modality of instrumentality, efficiency and materiality, the distinction between technological means and moral ends subsides, and technologies and moralities emerge as inherently mingled.

The notion that technologies help shape the basis of moral decisions, that behavior resulting from technological mediation can be understood as moral action, and that moral decision-making needs to be understood as a hybrid effort in which both humans and non-humans take part, requires a rethinking of the status of subjects as well as objects in ethical theory. I have argued that methodological posthumanism lacks this inquiry into subjectivity, focusing in most part on objects or non-humans, rather than on humans. But, as Verbeek (2011) argues, without such a rethinking of subjectivity in relation to technological mediation, there is a risk that the effect of technology's participation in this interplay will continue to be perceived as no more than an external limiting force on human freedom and morality. Therefore, a deeper understanding of the subject in relation to technological mediation is needed. And Foucault's work on ethics can be helpful here. I reframe this here in terms of technologically mediated subjectivity and genetic responsibility.

8.2.3   Genetic Technologies of the Self

As discussed earlier, Foucault's work on subjectivity can be roughly divided into two periods, the major part of his works, culminating in *Discipline and Punish* (1979), which concentrates on the historical research into the conditions of formation of modes of subjectivity, by and large as the result of disciplinary power, and his later works (1985, 1986, 1997a, b), which emphasize the practices of experimentation with or transformation of new modes of existence. This division can also be understood as a shift in focus from subject formation as an explicitly coercive and repressive process to subject formation as a productive and aesthetic process that individuals can in part take upon themselves. In this shift, a new space of relative freedom in relation to the government and stylizing of one's own existence opens up. In this later period, Foucault seeks to overcome the modern approach to ethics as a principle-based system by exploring the notion of ethics developed in classical Antiquity, in which ethics is directed at constituting oneself as a specific subject. In this framework, ethics is the explicit shaping of one's subjectivity by the "subjection" of oneself to a moral code, i.e. to a specific understanding of what constitutes the good life. As we have seen, what Foucault calls "technologies of the self", those practices that take one's body, thoughts and conduct as a site for work, in the aim of transforming oneself into a specific moral individual, should be seen as ethical practices. In his most recent work, Peter-Paul Verbeek (2008a, b, 2011) argues that such an approach to ethics in terms of moral self-constitution is particularly relevant for the ethics of technology because it integrates both the notion that the subject is technologically mediated – that is, an understanding that humans and technology cannot be viewed as separate – *and* the idea that the subject can actively relate to, or help to shape, these mediations. It is this last element that is of most significance for understanding the genetically responsible subject.

For Foucault, ethics involves being able to reflect on the processes by which we are constantly being constituted, transformed and shaped as subjects. Subjects are not only the result of a repressive disciplinary power, but can also develop an active relationship to the mediations, to the technologies of the self, that help shape them, by acknowledging, understanding and acting upon them, and by making them the object of the activity of subject constitution. New genetic technologies and the language of genetic risk can also be seen as providing an ethical framework that informs decisions on how to conduct one's life and allows individuals to develop a relationship to, and actively intervene in, the technological mediations that help shape their subjectivity. As Verbeek argues, ethics in this kind of perspective is not about "protecting" humanity from the threat of technology, but about explicitly shaping technological mediations, taking advantage of technology, in order to form the way in which we are constituted as subjects in a desirable manner. The knowledge derived from genetic technologies reconfigures identity in terms of a genetic present and a genetic future according to new values about obligations and hopes. These technologies can thus be seen as "genetic technologies of the self": they identify and problematize aspects of the self to be worked on, they elaborate means for managing them and for aiming at certain forms of life. As Rose (2007) argues, they imply an obligation to live one's life as a project, and a mode of subjectivity framed in terms of self-actualization, choice, and responsibility.

The new medical genetics, whether in the form of clinical counseling and surveillance or preventive medical intervention, shape individuals as specific moral subjects and create new ethical responsibilities. These responsibilities are not just directed at oneself in a narrow, individualizing sense, but to all those who share, or might share in the future, one's genome. Genetic individuality thus also has a collectivizing moment, since families, more than individuals, are the patients of genetic services (Richards 1996). In order to obtain information about one's genetic risk, information is needed about kin, and any information one obtains about oneself also has implications for those family members.[15] By definition then, genetic risk is never an individual risk, it is always shared with biological relatives. The experience of genetic testing itself may alter one's relationships with family and kin, causing a shift in the way individuals think about themselves in relation to others. This is to say that while the notions of freedom of choice, the right to know, informed decision-making and self-determination make up the normative basis of genetic testing, this new sense of responsibility may express a shift away from this individuated, autonomous self, to a more "interdependent" self that is constructed in relation to the needs of others (Kenen 1994). Nina Hallowell (1999), for example, in her study of how women identified as potentially at risk of developing hereditary breast and ovarian cancer perceived their choices, concludes that they did not only view themselves as responsible for their own and significant others' genetic risks, but also, as the significant others of others, felt an obligation towards others to manage these risks and to persuade others to act upon that information. For women who attended genetic counseling, this was seen as a first step towards taking responsibility for their risks and was justified, even if there would have been an individual preference "not to know", in terms of the implications for other family members. As Hallowell remarks, "genetics is not about individuals, it is about biological relationships" (1999: 106).

And biological – in what seems to be a reversal of trends we saw earlier with reproductive medicine – need not necessarily mean kin. The interdependent character of genetic subjectivity also implies that individuals sharing confirmed or potential genetic risks become part of complex networks of social relatedness that can carry a set of obligations of care. Individuals who have tested positive for genetic markers sometimes volunteer to share the results of genetic testing in order to contribute to research on the genetics of their potential disease. In some research on genetic

---

[15] This has led to extensive debate on the right (not) to know of a patient's relatives (see for example Chadwick 1997; Laurie 1999).

technologies, for example, altruism has been cited by participants as a main motivation for participating in clinical research (Kerr et al. 1998). In other cases, individuals volunteer their genetic data towards the creation of public databases. The "Personal Genome Project" for example, launched by the Harvard Medical School, aims to publish the complete genomes of its volunteers along with extensive information about their phenotype (medical records, measurements, MRI images, etc.), which will all be freely available to researchers and the public on the Internet. By using real medical and genomic data from specific people, the project also aims at creating test cases for the legal and ethical issues surrounding the availability of personal and genomic records, giving ethicists, legislators, and scientists concrete examples to study. These new forms of genetic responsibility can be related to what Rayna Rapp (1999) has termed "moral pioneers". Like those couples who were the first to face the complex decisions generated by the technology of amniocentesis, these individuals can be seen as pioneering a new informed ethics of the self.

8.2.4 Genetically Responsible Collectives

Genetic responsibility also makes it possible to discern new forms of social identity and political participation in the form of support groups, online communities and activist collectivities. The idea of "biosociality" (Rabinow 1996), "genetic citizenship" (Heath et al. 2004; Petersen 2002) or "biocitizenship" (Rose 2005) indicates how genetic, or more generally biological, categories are increasingly forming the basis for belonging to certain communities. This new form of biocitizenship, argues Rose, is manifested

> in a range of struggles over individual identities, forms of collectivization, demands for recognition, access to knowledge, and claims to expertise. It is generating new objects of contestation, not least those concerning the respective powers and responsibilities of public bodies, private corporations, health providers and insurers, and individuals themselves. It is creating novel forums for political debate, new questions for democracy, and new styles of activism. (2007: 136–137)

By invoking the term citizenship, this notion also emphasizes how genetically responsible subjectivity is not confined to individuality, but is associated with the rise of patient advocacy and health activism. The formation of collectivities around a biological conception of shared identity in itself is not recent. But these collectivities are playing an ever greater role as stakeholders, in actively influencing how research is conducted, and ensuring that patient perspectives are taken into account by institutional actors (Rabinow 1999; Callon 2008; Novas 2006).[16]

---

[16] In his work on people affected by muscular dystrophy in France in the 1990s, Rabinow (1999) identified a shift in the ways patients and families relate to their disease, observing that they abandon more traditional models of support for an active and mobilized attitude characterized by a collaboration with researchers (donation of blood samples) and funding of genomic research.

The Internet has become a powerful new means for the formation of biociti-zenship, as both a vast source of medical information and as the ground upon which bio-communities can proliferate and grow. Websites, chat rooms and forums can now be found for almost every type of known disease and genetic disorder, or as search engines for them. Web communities are networks where subjects at risk, patients and caregivers can share their experiences of what it means to live with an illness, how to manage treatment side effects, how to negotiate access to health care, where to find more information, and how to actively influence research or policy (see for example, "PatientsLikeMe", "CureTogether" "Genetic Alliance" and "FORCE" ["Facing Our Risk of Cancer Empowered"]). These communities express the quintessence of the configuration of a new ethics of biocitizenship and genetic responsibility that is emerging from these novel technological mediations that subjects are engaged in today.

On all of the websites, the narratives of diagnosis, coping and management of confirmed or potential disease, in the forms of individual stories, postings and comments, create a language via which veterans and newcomers reflect upon and construct their genetic subjectivity, including specific jargon for each commu-nity. Just as Foucault understood the practices of confession and diary writing as technologies of the self, so do these practices help shape individuals as ethical subjects (Novas 2003).[17] Individual blogs are also a common practice of indi-viduals living with disease today. Here too, the disclosure of one's experience can be both a source for others to identify themselves with and the constitution, by means of self-reflection, of one's own genetic or biological identity. A term that has become an integral part of FORCE's lexicon, for example, and has spread to the medical and research community, is "cancer pre-vivor", to desig-nate a "survivor of a predisposition to cancer … [such as] people who carry a hereditary mutation, a family history of cancer, or some other predisposing factor and who are living with the knowledge of being high-risk". This term attests to both the individualizing and collectivizing moments of the constitution of the genetic mode of subjectivity, via which new types of group and individual identi-ties are arising out of the new techniques of genetic diagnosis and monitoring of risks.

---

[17] Foucault claims:

Writing was also important in the culture of the care of the self. One of the tasks that defines the care of the self is that of taking notes on oneself to be reread, writing treatises and letters to friends to help them, and keeping notebooks in order to reactivate for oneself the truths one needed … Taking care of oneself became linked to constant writing activity. The self is something to write about, a theme or object (subject) of writing activity. (1997b: 232)

8.2.5 The Critique of Genetic Responsibility: From Empowerment to Discipline

In the context of genetic risk and biocitizenship, individuals often engage with the management of their genetic risks as consumers of medical resources, driven by a right to know, a right to access and a right to decide. Indeed, genetic responsibility is often framed in a rhetoric that echoes liberal posthumanist discourse, of empowerment, autonomy and choice. This convergence of consumer market logic and health discourses has been the basis for a great deal of criticism of the new genetic technologies. It has stirred numerous warning calls about idealized images of empowered individuals making free, informed choices in an unregulated genomic marketplace from many sides of the academic spectrum, from dystopic posthumanists (Fukuyama 2004; Sandel 2004), to radical posthumanists (Haraway 1997) and sociologists of health and illness (Kerr and Shakespeare 2002; Nettleton 1997; Petersen and Bunton 2002). Here theorists are often concerned with how taking on genetic responsibility is less a matter of choice than of a moral obligation to do so. Responsibility, in this sense, can be more constraining than it is empowering. In the context of antenatal screening (Katz Rothman 1994), for example, while there is no mandatory obligation to undergo screening, there are clear recommendations expressed by medical authorities to do so, that carry with them quite explicit presuppositions that bringing to life a baby with Down's Syndrome should be avoided at all costs.[18] Or, in the context of genetic testing for inheritable forms of cancer, as Hallowell (1999) argues, individuals often do not experience what we would call a "free" choice, because their genetic risk, as we have seen, entails responsibilities to others.

Thus the decision to obtain genetic information and how to act upon it is frequently influenced by third parties in a way that problematizes notions of personal choice and empowerment – from medical experts and their portrayal of what positive health behavior is (Johanson 2000; Williams et al. 2002), to the interests of other family members, and broader societal expectations and norms of what responsible conduct means. This has implications for the very notion of "informed consent", since every act of consent must be considered against a backdrop of all of the potential outcomes of refusing consent: one is responsible for one's biomedical condition, both for better and for worse, and those who choose not to adopt an informed and prudent relation to their genetic selves easily become the subject of social stigma.[19] This is to say that the constitution of genetically "responsible" subjectivity is always accompanied by the constitution of genetically "irresponsible" subjects (Callon and Rabeharisoa 2004).

These concerns are extremely problematic as to my claim that the emergence of a genetically responsible mode of subjectivity not only eludes fears of genetic determinism and essentialism, but that it is a non-humanist technologically mediated

---

[18] In the mid-1990s, approximately 90 % of fetuses with Down's syndrome were aborted in France, the US and the UK. See Mansfield et al. (1999) for an overview.

[19] As in the debate around personal responsibility for "lifestyle diseases" like obesity.

mode of subjectivity that can act as an alternative to the liberal posthumanist model of a technologically-enhanced rational autonomous subject. Genetic responsibility in these critiques is associated with neoliberal models of government and "entrepreneurial" rationalities of the self, characterized by a withdrawal of the state from the public domain and a transfer of responsibilities to pro-active, involved citizens who can care for themselves. Petersen writes:

> Neo-liberalism calls upon the individual to enter into the process of their own self-governance through the processes of endless self-examination, self-care, and self-improvement. In other words, the entrepreneurial subject is reconceived – and reproduced – as a new kind of citizen: a neo-liberal citizen who is autonomous, responsible and self-governing. (Petersen and Lupton 1996: 51)

It is not entirely clear in this sense how genetically responsible selfhood fundamentally differs from the narrative of liberal humanist selfhood, with its capacity for self-control, mastery and rationality, that a mediated type of non-humanist posthumanism aspires to deconstruct. The rhetoric of empowerment and enablement that often accompanies this mode of subjectivity can be seen as constructing subjects as "health consumers" in accordance with the model of consumer capitalism. Indeed, the link is easily made between the notion of self-care that underlies genetic responsibility and liberal humanist values expressed in the possibilities for individuals to be in control over their destinies and to have the capacity for self-control, responsibility, rationality and enterprise (Nettleton 1997). Claims of empowerment are seen in this context as a tool that enables a transfer of responsibilities from the state to the individual. And genetically responsible subjects are construed as subjects who have internalized the normalizing health discourse of neo-liberal regimes – as subjects that have been disciplined rather than empowered.

Foucault's (2008) work on governmentality and individualized forms of self-regulation in the constitution of modern subjectivities has been very influential in this type of critique. Here modern techniques of health promotion are seen as a form of government that oversees the "proper routes to health" through a disciplining of the self (Coveney 1998: 462). This process, it is argued, in turn assists the modern state's surveillance and control of human populations, so that it has an important biopolitical component, as we have seen. I would like to suggest however, that this Foucauldian-based critique of health promotion and genetically responsible subjectivity, while clearly important when attempting to critically assess the impact of new genetic technologies, misses important aspects of genetically responsible subjecthood, and that this may proceed precisely from a narrow reading of Foucault's work on ethics.

8.2.6   Freedom as Non-Sovereignty

Implicit in this critique is the idea that genetic responsibility and the rhetoric of individual empowerment that often surrounds it actually disguises a *lack* of freedom that results from the obligation to engage in personal risk management and a duty to behave responsibly – as contrived by the state or the market place. In these terms, for example, the "new eugenics" expresses the concern that individuals may believe they are acting in accordance with independently conceived choices, when really, they are merely carrying out the imperatives of hegemonic health and consumer trends that have been internalized.[20] Thus it is the apparent but unidentified lack of autonomy on the part of genetically responsible subjects that is the focus of this critique. Genetically responsible subjects have been duped, so to speak, on various levels, unwittingly contributing to processes that are not in their best interest, from the privatization of health to new forms of "free labor" (Harris et al. 2010).

Viewed from the perspective of technological mediation, however, the concept of freedom that informs this critique presupposes a form of sovereignty with respect to technology that human beings simply do not possess. That is, technological mediation already recognizes that human beings are not fully autonomous in their subject constitution, so that the contention concerning a lack of freedom or sovereignty here is in itself misplaced. This does not mean that the notion of freedom dissolves in the framework of technological mediation. On the contrary, departing from the reading of Foucault as presented in Chap. 6, freedom becomes a prerequisite of the ethical framework in which genetically responsible subjects emerge. As we have seen, in his later writings, Foucault's works are characterized by two seemingly disparate projects, his examination of the formation of the state, or technologies of domination, and a focus on ethical questions and the constitution of the subject, or technologies of the self. It is the problem of government, as the "conduct of conduct", and by extension of freedom, that connects these two research interests. In this sense, just as it is important to take into account both the coercive and the productive aspects of subject formation in Foucault's work, technologies of domination, or the government of subjects through freedom, must be understood alongside technologies of the self, or subject constitution as an ascetic practice that one applies to oneself in a space of *relative* freedom. Foucault writes:

> I think that if one wants to analyze the genealogy of the subject in Western civilization, he has to take into account not only techniques of domination but also techniques of the self. … to take into account the points where the technologies of domination of individuals over one another have recourse to processes by which the individual acts upon himself. And conversely, he has to take into account the points where the techniques of the self are integrated into structures of coercion and domination. The contact point, where the individuals are driven by others is tied to the way they conduct themselves, is what we can call, I think, government. Governing people, in the broad meaning of the word, governing people is not a way to force people to do what the governor wants; it is always a versatile equilibrium, with complementarity and conflicts between techniques which assure coercion and processes through which the self is constructed or modified by himself (1993: 203–204)

Foucault maintains that the ethics of caring for oneself is a way to exercise freedom, that ethics *is* the conscious practice of freedom (see especially Foucault 1997a).

---

[20] Sandel (2004: 9) writes:

> What, after all, is the moral difference between designing children according to an explicit eugenic purpose and designing children according to the dictates of the market? Whether the aim is to improve humanity's "germ plasm" or to cater to consumer preferences, both practices are eugenic.

But, in light of Foucault's historicization of questions of ontology, in which there are no essences but only the forging of identities through processes of self-formation, the understanding of freedom as an exercise of "liberation" of a true self from bondage or repression, or the understanding of freedom as some absolute lack of all constraint or the absence of "external" influences, makes no sense. The understanding of freedom as the ontological condition for ethics in Foucault's thought implies that freedom takes the form of a kind of informed reflection in the ongoing self's relationship to itself, in the ethical work that a person performs on their self. As I have argued, freedom here is the ability to relate to precisely those "external" influences that determine the self. It exists in the possibilities that human beings have to help shape their relationship to the material environment in which they live and to which they are bound (Verbeek 2011: 9). And it is the understanding that one is never an entirely sovereign, independent being – that there can be no such thing, that opens up a space of freedom.

In terms of our technological culture, and the emergence of a technologically mediated, genetically responsible subjecthood, this understanding implies that the engagements with new genetic technologies can be a site of freedom rather than an absence of it. In the "versatile equilibrium" that Foucault insists upon, between technologies of domination and technologies of the self, the productive and creative practices of subject constitution must be identified and encouraged. But the Foucauldian critique of health promotion seems to omit this important point, by placing too much emphasis on the disciplinary and repressive – even though not necessarily directly coercive – aspect of subject constitution, and overlooking many of the creative and productive aspects of genetically responsible subjectivity. This critique can prevent us from seeing how many engagements with new genetic technologies resist disciplinary narratives and express a creativity and resourcefulness on the part of users that should be identified as an active appropriation or a "stylizing" of the technological mediations that contribute to the constitution of the self.

Thus, a growing number of studies that follow-up on what individuals do with the information that is obtained from genetic testing, both in the medical and nonmedical context, and what kind of impact this information has, indicate some unexpected findings. For example, in a large study on the psychological and behavioral impact on individuals considered to be at risk for Alzheimer's disease, Chilibeck et al. (2011) found that participants who were counseled about their personal genetic risk for Alzheimer's tended to combine the genetic information they received with existing ideas about heredity, fusing these quite incompatible discourses in various ways. While this conflation could be interpreted as a "misunderstanding" of the science, the researchers argue that it should rather be seen as a means of expressing abstract and complex information in concrete and familiar terms. In this process, genetic information is absorbed, selectively used and integrated, or "familiarized" differently by different people. This should be seen as a creative process of identity formation in a context of genetic risk knowledge.

Arribas-Ayllon et al. (2011), in an ethnographic study that looks into the reasons for the *non-disclosure* of genetic risk between members of families who attended a clinical genetic service, also argue that genetic knowledge is not translated into new patterns of obligation – of risk disclosure in this case – in any straightforward way. They found that the neoliberal orientation towards risk management that underlies a duty to disclose risk information is only one of the vectors at play in this context, alongside other, no less influential ones, including mundane understandings of inheritance and structures of kinship. Namely, the families in their study found innovative ways for managing the fear that is involved in the disclosure of genetic risk to family members, both the fear of negatively affecting others and the fear of being blamed for carrying that information. Non-disclosure should be interpreted here not as an expression of a failure to fulfill one's duty, but as a way of defending family members, including oneself, from the impact of genetic knowledge. It is a "clandestine strategy of familial preservation in which, unlike the preventive logic of healthism, pursues a practical logic of surveillance and protectionism" (2011: 20). This implies a different sense of genetic responsibility, one that is not framed in terms of rational autonomy.

Biological conceptions of personhood and genetic identity are rarely hegemonic. As Novas and Rose (2000) argue, attributions of genetic risk will rewrite personhood in "unexpected ways". This is not just because it seems quite impossible to predict what behavioral impact the knowledge produced via genetic technologies will lead to. But more importantly, individuals engage with the knowledge produced by genetic testing in strategic ways to actively construct their own identities. Genetic knowledge production is explicitly involved in mediating subject constitution. This poses a serious challenge to claims of biological determinism and geneticization, the idea that subjects are increasingly understanding themselves and social phenomena solely in terms of the biological, as well as fears that new genetic technologies will act as sites for biopolitical control via the implementation of specific types of healthy conduct. In the context of genetic responsibility, as in the contexts of assisted reproduction and neuroscience, biological knowledge seems to be mobilized and appropriated as part of one's self-understanding in an unforeseeable way.

8.3 Conclusion

At the intersection of biology's determinedness and biology's manipulability, of nature's givenness and nature's perfectibility, a new mode of subjectivity that implies moral duties and new fields of responsible action is emerging in the form of genetically responsible selfhood. This new understanding of personhood relates to a number of other contemporary phenomena, such as the introduction of the discourse of genetic risk, the individualization and privatization of social risks, and a growing emphasis on norms of entrepreneurship and self-actualization characteristic of liberal democratic societies. This mutation in personhood is an effect of geneticization, the process by which human biology and identity are increasingly being explained by, understood and framed in genetic terms. But genetically responsible subjectivity implies precisely the surplus within the notion of geneticization that critiques of genetic determinism and essentialism, of growing medicalization and the extension of biopolitics into the very biology of life itself, cannot account for. The new genetic technologies do not generate resignation, fatalism or passivity in the face of biological determinism. On the contrary, they create an obligation to act in the present by adopting an informed and prudent relation to the future. Thus the genetically responsible subject emerges within a framework of new obligations towards moral decision-making, self-actualization and responsibility to others, and the understanding that one can have an *active* relationship with the technological mediations that help constitute the self.

Such critiques are thus of little help for shedding light on the profound repercussions new biotechnologies have on subjectivity, and their skepticism and pessimism prevent them from identifying many of the positive and enriching effects of these technologies – not just for the realm of medical treatment, but also for what it means to be human. Likewise, an understanding of genetic governmentality as a mere expression of market capitalism in the context of health consumption, while more intricate than critiques of genetic determinism, also fails to grasp the profound implications that technological mediation – in which the genetically responsible mode of subjectivity is immersed – has for notions of freedom and autonomy. What is needed is a mediated posthumanist approach which can incorporate both the non-humanist notion that the subject is technologically mediated, i.e., an understanding that humans are in part constituted by their technologies, and the understanding that the subject can actively relate to, or help shape these mediations. An ethics of technology that can offer a framework for assessing emerging enhancement and biotechnologies in this perspective is not about protecting humanity from the threat of technology, but about explicitly shaping technological mediations, taking advantage, so to speak, of technology, in order to shape the way in which we are constituted as subjects in a desirable manner.

References

Alford, J., et al. (2005). Are political orientations genetically transmitted? *American Political Science Review, 99*(2), 153–167.

Armstrong, D. (2008). The rise of surveillance medicine. *Sociology of Health and Illness, 17*(3), 393–404.

Arribas-Ayllon, M., Featherstone, K., & Atkinson, P. (2011). The practical ethics of genetic responsibility: Non-disclosure and the autonomy of affect. *Social Theory & Health, 9*(February), 3–23.

Bains, W. (2001). The parts list of life. *Nature Biotechnology, 19*, 401–402.

Bartels, A., & Zeki, S. (2000). The neural basis of romantic love. *NeuroReport, 1*(17), 3829–3834.

Beck, U. (1992). *Risk society: Towards a new modernity*. London: Sage.

Bouchard, T. J., Lykken, D. T., McGue, M., Segal, N. L., & Tellegen, A. (1990). Sources of human psychological differences: the minnesota study of twins reared apart. *Science, 250*(4978), 223–228.

Brenninkmeijer, J. (2010). Taking care of one's brain: How manipulating the brain changes people's selves. *History of the Human Sciences, 23*(1), 107–126.

Bunton, R. (1997). Popular health, advanced liberalism and good housekeeping. In A. Petersen & R. Bunton (Eds.), *Foucault, health and medicine* (pp. 223–248). New York: Routledge.

Callon, M. (2008). The growing engagement of emergent concerned groups in political and economic life: Lessons from the French Association of Neuromuscular Disease Patients. *Science, Technology and Human Values, 33*(2), 230–261.

Callon, M., & Rabeharisoa, V. (2004). Gino's lesson on humanity: Genetics, mutual entanglements and the sociologist's role. *Economy and Society, 33*(1), 1–27.

Caspi, A., Sugden, K., Moffit, T. E., Taylor, A., Craig, I. W., Harrington, H. L., McClay, J., Mill, J., Martin, J., Braithwaite, A., & Poulton, R. (2003). Influence of life stress on depression: Moderation by a polymorphism in the 5-HTT gene. *Science, 301*(5631), 386–389.

Castel, R. (1981). *La gestion des risques, de l'anti-psychiatrie à l'après-psychanalyse*. Paris: Minuit.

Chadwick, R., Levitt, M., & Shickle, D. (1997). *The right to know and the right not to know.* Aldershot: Averbury.

Chilibeck, G., Lock, M., & Sehdev, M. (2011). Postgenomics, uncertain futures, and the familiarization of susceptibility genes. *Social Science and Medicine, 72*(11), 1768–1775.

Collins, F. S., Lander, E. S., Rogers, J., & Waterson, R. H. (2004). Finishing the euchromatic sequence of the human genome. *Nature, 431*(7011), 931–945.

Conrad, P. (2007). *The medicalization of society: On the transformation of human conditions into treatable disorders*. Baltimore: Johns Hopkins University Press.

Costa, P. T., & McCrea, R. R. (1992). *Revised NEO personality inventory (NEO-PI-R) and NEO five-factor inventory (NEO-FFI) manual*. Odessa: Psychological Assessment Resources.

Coveney, J. (1998). The government and ethics of health promotion: The importance of Michel Foucault. *Health Education Research, 13*(3), 459–468.

Crick, F. (1958). On protein synthesis. *Symposia of the Society for Experimental Biology, 12*, 138–163.

Dawkins, R. (1989). *The selfish gene*. Oxford: Oxford University Press.

Dawkins, R. (1999). *The extended phenotype: The long reach of the gene*. Oxford: Oxford University Press.

Deleuze, G., & Guattari F. (1977). *Anti-Oedipus: Capitalism and schizophrenia* (trans: Seem, M., Lane, H.R., & Hurley, R). New York: Viking Press. Original edition, 1972.

Deleuze, G., & Guattari, F. (1987). *A thousand plateaus: Capitalism and schizophrenia* (trans Massumi, B.). Minneapolis: University of Minnesota Press. Original edition, 1980.

Dillon, M. (2000). Poststructuralism, complexity and poetics. *Theory, Culture and Society, 17*(5), 1–26.

Doyle, R. (1997). *On beyond living: Rhetorical transformations of the life sciences*. Stanford: Stanford University Press.

Elliot, C. (2003). *Better than well: American medicine meets the American dream*. New York: W.W. Norton & Company.

Foucault, M. (1979). *The history of sexuality, volume 1: An introduction* (trans: Hurley, R.). London: Allen Lane. Original edition, 1976.

Foucault, M. (1985). *The use of pleasure* (trans: Hurley, R.). New York: Pantheon. Original edition, 1984.

Foucault, M. (1986). *The care of the self* (trans: Hurley, R.). New York: Pantheon. Original edition, 1984.

Foucault, M. (1989). *The order of things: An archaeology of the human sciences*. London/New York: Pantheon. Original edition, 1966.

Foucault, M. (1993). About the beginning of the hermeneutics of the self: Two lectures at Dartmouth. *Political Theory, 21*(2), 198–227.

Foucault, M. (1997a). The ethics of the concern for self as a practice of freedom. In P. Rabinow (Ed.), *Ethics, subjectivity and truth: The essential works of Michel Foucault 1954–1984, vol. 1* (pp. 281–301). New York: The New Press. Original edition, 1994.

Foucault, M. (1997b). Technologies of the self. In P. Rabinow (Ed.), *Ethics, subjectivity and truth: of the essential works of Michel Foucault 1954–1984, vol. 1* (pp. 223–251). New York: The New Press. Original edition, 1994.

Foucault, M. (2008). *The birth of biopolitics: Lectures at the Collège de France* (trans: Burchell, G.). New York: Palgrave. Original edition, 2004.

Franklin, S. (1993). Essentialism, which essentialism? Some implications of reproductive and genetic technoscience. *Journal of Homosexuality, 24*(3–4), 27–40.

Franklin, R., & Gosling, R. (1953). Molecular configuration in sodium thymonucleate. *Nature, 171*, 740–741.

Franklin, S., & Roberts, C. (2006). *Born and made: an ethnography of preimplantation genetic diagnosis*. Princeton: Princeton University.

Fraser, M. (2001). The nature of Prozac. *History of the Human Sciences, 14*(3), 56–84.

Fukuyama, F. (2004). The world's most dangerous ideas. *Foreign Policy, 144* (September/October 2004), 32–33.

Golberg, L. R. (1993). The structure of phenotypic personality traits. *American Psychologist, 48*, 26–34.

Greenfield, S. (2000). *The human brain: A guided tour*. London: Phoenix.

Griggers, C. (1997). *Becoming-woman*. Minneapolis/London: University of Minnesota Press.

Grosz, E. (1994). *Volatile bodies: Towards a corporeal feminism*. Bloomington: Indiana University Press.

Hallowell, N. (1999). Doing the right thing: Genetic risk and responsibility. In P. Conrad & J. Gabe (Eds.), *Sociological perspectives on the new genetics*. Oxford: Blackwell.

Haraway, D. (1997). *Modest_Witness@Second_Millenium. FemaleMan©_Meets_Oncomouse™: Feminism and Technoscience*. New York: Routledge.

Harris, R., Wathen, N., & Wyatt, S. (2010). *Configuring health consumers: Health work and the imperative of personal responsibility*. Basingstoke: Palgrave Macmillan.

Heath, D., Rapp, R., & Taussig, K.-S. (2004). Genetic citizenship. In D. Nugent & J. Vincent (Eds.), *Companion to the anthropology of politics* (pp. 152–167). Oxford: Blackwell.

Horwitz, A., & Wakefield, J. (2007). *The loss of sadness: How psychiatry has transformed normal sadness into depressive disorder*. Oxford: Oxford University Press.

Howson, A. (2004). *The body in society*. Cambridge: Polity Press.

Hubbard, R., & Wald, E. (1997). *Exploding the gene myth*. Boston: Beacon.

Illich, I. (1975). *Limitis to medicine. Medical nemesis: The expropriation of health*. London: Marion Boyars.

Jacob, F. (1973). *The logic of life: A history of heredity* (trans: Spillmann, B.E.). New York: Pantheon. Original edition, 1970.

Jameson, F. (1984). Postmodernism, or the cultural logic of late capitalism. *New Left Review, 146*(1), 53–92.

Jaroff, L. (1989). The gene hunt. *Time*, March 20, 62–67.

Johanson, R., Burr, R., Leighton, N., & Jones, P. (2000). Informed choice? Evidence of the persuasive power of professionals. *Journal of Public Health Medicine, 22*(3), 439–440.

Katz Rothman, B. (1994). *The tentative pregnancy: Amniocentesis and the sexual politics of motherhood*. London: Pandora.

Kay, L. E. (2000). *Who wrote the book of life? A history of the genetic code*. Stanford: Stanford University Press.

Keller, E. F. (1995). *Refiguring life: Metaphors of twentieth century biology*. Colombia: Colombia University Press.

Keller, E. F. (2000). *The century of the gene* (3rd ed.). Cambridge, MA: Harvard University Press.

Kenen, R. (1994). The Human Genome Project: Creator of the potentially sick, potentially vulnerable and potentially stigmatized? In I. Robinson (Ed.), *Life and death under high technology medicine* (pp. 49–64). Manchester: Manchester University Press.

Kerr, A., & Shakespeare, T. (2002). *Genetic politics: From eugenics to genome*. Cheltenham: New Clarion Press.

Kerr, A., Cunningham-Burley, S., & Amos, A. (1998). The new genetics and health: Mobilizing lay expertise. *Public Understanding of Science, 7*, 41–60.

Kevles, D. (1995). *In the name of eugenics*. Cambridge, MA: Harvard University Press.

Kleinman, A. (1988). *Rethinking psychiatry: From cultural category to personal experience*. New York: Free Press.

Kramer, P. D. (1993). *Listening to Prozac: A psychiatrist explores antidepressant drugs and the remaking of the self*. New York: Penguin.

Latour, B. (1987). *Science in action: How to follow scientists and engineers through society*. Cambridge: Harvard University Press.

Latour, B. (1992). Where are the missing masses? Sociology of a few mundane artifacts. In W. E. Bijker & J. Law (Eds.), *Shaping technology/building society: Studies in sociotechnological change* (pp. 225–259). Cambridge, MA: MIT Press.

Latour, B. (1994). On technical mediation: Philosophy, sociology, genealogy. *Common Knowledge, 3*, 29–64.

Latour, B. (2002). Morality and technology: The end of the means. *Theory, Culture and Society, 19*(5–6), 247–260.

Laurie, G. (1999). In defence of ignorance: Genetic information and the right not to know. *European Journal of Health Law, 6*, 119–132.

Lemke, T. (2002). Genetic testing, eugenics and risk. *Critical Public Health, 14*(3), 49–64.

Lemke, T. (2004). Disposition and determinism: Genetic diagnostics in risk society. *The Sociological Review, 52*(2004), 550–566.

Lemke, T. (2005). "A zone of indistinction": A critique of Giorgio Agamben's concept of biopolitics. *Outlines. Critical Social Studies, 7*(1), 3–13.

Lemke, T. (2011). *Biopolitics. An advanced introduction*. New York: New York University Press.

Lewontin, R. C. (1991). *Biology as ideology: The doctrine of DNA*. New York: Harper Perennial.

Lippman, A. (1992). Led (astray) by genetic maps: The cartography of the human genome and health care. *Social Science and Medicine, 35*(12), 1469–1476.

Lippman, A. (1993). Prenatal genetic testing and geneticization: Mother matters for all. *Fetal Diagnosis and Therapy, 8*, 175–188.

Lykken, D. (1999). *Happiness: The nature and nurture of joy and contentment*. New York: St. Martin's Press.

Lykken, D., & Tellegen, A. (1996). Happiness is a stochastic phenomenon. *Pyschological Science, 7*(3), 186–189.

Mansfield, C., Hopfer, S., & Marteau, T. M. (1999). Termination rates after prenatal diagnosis of Down syndrome, spina bifida, anencephaly, and Turner and Klinefelter syndromes: A systematic literature review. *Prenatal Diagnosis, 19*(9), 808–812.

Martin, E. (2010). Self-making and the brain. *Subjectivity, 3*(4), 366–381.

Masters, R. D., & McGuire, M.T., (Eds.), (1994). *The neurotransmitter revolution: Serotonin, social behavior, and the law*. Carbondale, Ill.: Southern Illinois University Press.

Murray, C., & Herrnstein, R. J. (1994). *The bell curve: Intelligence and class structure in American life*. New York: Free Press.

Nelkin, D., & Lindee, M. S. (1995). *The DNA mystique: The gene as a cultural icon*. New York: W.H. Freeman and Company.

Nelkin, D., & Tancredi, L. (1994). *Dangerous diagnostics: The social power of biological information* (2nd ed.). Chicago: University of Chicago Press.

Nettleton, S. (1997). Governing the risky self: How to become healthy, wealthy and wise. In A. Petersen & R. Bunton (Eds.), *Foucault, health and medicine* (pp. 207–222). London: Routledge.

Novas, C. (2003). *Governing risky genes*. Dissertaion, University of London, London.

Novas, C. (2006). The political economy of hope: Patients' organizations, science and biovalue. *Biosocieties, 1*(3), 289–305.

Novas, C., & Rose, N. (2000). Genetic risk and the birth of the somatic individual. *Economy and Society, 29*(4), 484–513.

Oyama, S. (2000). *The ontogeny of information: Developmental systems and evolution*. Durham: Duke University Press.

Parens, E. (1998). Is better always good? *Hastings Center Report, 28*(1), S1–S17.

Petersen, A. (1996). Risk and the regulated self: The discourse of health promotion as politics of uncertainty. *Australian New Zealand Journal of Sociology, 31*(1), 44–57.

Petersen, A., & Bunton, R. (2002). *The new genetics and the public's health*. London: Routledge.

Petersen, A., & Lupton, D. (1996). *The new public health: Health and self in the age of risk*. London: Sage.

Pickersgill, M., Cunningham-Burley, S., & Martin, P. (2011). Constituting neurologic subjects: Neuroscience, subjectivity and the mundane significance of the brain. *Subjectivity, 4*(3), 346–365.

Rabinow, P. (1992). Artificiality and enlightenment: From sociobiology to biosociality. In J. Crary & S. Kwinter (Eds.), *Incorporations* (pp. 234–252). New York: Zone.

Rabinow, P. (1996). *Essays on the anthropology of reason*. Princeton: Princeton University Press.

Rabinow, P. (1999). *French DNA: Trouble in purgatory*. Chicago: University of Chicago Press.

Rabinow, P., & Rose, N. (2006). Biopower today. *Biosocieties, 1*(2), 195–217.

Rapp, R. (1999). *Testing women, testing the fetus: The social impact of amniocentesis in America*. New York: Routledge.

Richards, M. (1996). Lay and professional knowledge of genetics and inheritance. *Public Understanding of Science, 5*, 217–230.

Risch, N., Herrell, R., Lehner, T., Liang, K.-Y., Eaves, L., Hoh, J., Griem, A., Kovacs, M., Ott, J., & Ries Merikangas, K. (2009). Interaction between the serotonin transporter gene (5-HTTLPR), stressful life events, and risk of depression. *The Journal of the American Medical Association, 301*(23), 2462–2471.

Rose, N. (1998). Governing risky individuals: The role of psychiatry in new regimes of control. *Psychiatry, Psychology and Law, 5*(2), 177–195.

Rose, N. (2007). *The politics of life itself: Biomedicine, power, and subjectivity in the twenty-first century*. Princeton: Princeton University Press.

Rose, N., & Novas, C. (2005). Biological citizenship. In A. Ong & S. Collier (Eds.), *Global assemblages: Technology, politics and ethics as anthropological problems* (pp. 439–463). Oxford: Blackwell.

Russell, M. T., & Karol, D. L. (1994). *The 16PF fifth edition administrator's manual*. Champaign, Il: Institute for Personality and Ability Testing.

Sandel, M. (2004). The case against perfection. *The Atlantic*, 1–11 April.

Savulescu, J. (2008). The perils of cognitive enhancement and the urgent imperative to enhance the moral character of humanity. *Journal of Applied Philosophy, 25*(3), 162–167.

Stahl, S. (1996). *Essential psychopharmacology: Neuroscientific basis and practical applications*. Cambridge: Cambridge University Press.

Stotz, K. C., Bostanci, A., & Griffiths, P. E. (2006). Tracking the shift to 'postgenomics'. *Community Genetics, 9*(3), 190–196.

Szasz, T. (1970). *The manufacture of madness*. New York: Dell.

Turkheimer, E. (2000). Three laws of behavior genetics and what they mean. *Current Directions in Psychological Science, 9*(5), 160–164.

Venter, J. C., et al. (2001). The sequence of the human genome. *Science, 291*(5507), 1304–1351.

Verbeek, P.-P. (2008a). Cultivating humanity: Toward a non-humanist ethics of technology. In E. Selinger, J. B. Olsen, & S. Riis (Eds.), *New waves in philosophy of technology* (pp. 241–263). Hampshire: Palgrave MacMillan.

Verbeek, P.-P. (2008b). Obstetric ultrasound and the technological mediation of morality: A post-phenomenological analysis. Human Studies, *31*(1), 11–26.

Verbeek, P.-P. (2011). *Moralizing technology: understanding and designing the morality of things*. Chicago: Chicago University.

Watson, J. D., & Crick, F. (1953a). Genetical implications of the structure of deoxyribonucleic acid. *Nature, 171*, 964–967.

Watson, J. D., & Crick, F. (1953b). A structure for deoxyribonucleic acid. *Nature, 171*, 737–738.

Williams, C., Aldersen, P., & Farsides, B. (2002). Is nondirectiveness possible within the context of antenatal screening and testing? *Social Science & Medicine, 54*(3), 339–347.

Wynne, B. (2005). Reflexing complexity: Post-genomic knowledge and reductionist returns in public science. *Theory, Culture and Society, 22*(5), 67–94.

Zeki, S., & Romaya, J. (2008). Neural correlates of hate. *PLoS One, 3*, e1128.

Chapter 9
Conclusion

Accounting for novelty is an extremely delicate affair. We should always be aware of our vulnerability to the assumption that the changes we are witnessing are the manifestation of, or the driving force behind, some fundamental break with the past, some radical epochal transformation. On the other hand, without the possibility to think and to identify the new, the fundamentally different, our reflective capacity as philosophers and social theorists, but also as self-conscious beings, would be greatly impaired. This makes the task of assessing forms of technological change and their implications for human experience very difficult. The four types of posthumanist discourse that were mapped out in this study offer four theoretical frameworks through which to view the implications of emerging biotechnologies for what it means to be human, through which to articulate what it is that strikes us as so novel in emerging biotechnologies. Each of them emphasizes a different aspect of the interaction between humans and technologies, each of which implies a certain understanding of human nature that comes to light in view of emerging biotechnologies. For dystopic posthumanism this relationship is framed in terms of transgression, and the human in terms of its givenness. For liberal posthumanism this relationship is framed in terms of mastery, and the human in terms of its transcendental aspirations. For radical posthumanism this relationship is framed in terms of deconstruction, and the human in terms of its prosthetic nature. And for methodological posthumanism this relationship is framed in terms of reciprocity, and the human in terms of its engagement with matter.

Throughout this study, I argued that each of these types of posthumanist discourse is limiting to varying degrees. Dystopic and liberal posthumanism, the two approaches that dominate the public debate surrounding emerging biotechnologies, even as they seem to stand in stark contrast, are both grounded in the humanist view that there exists a rigid separation between human beings and their technologies; a division that emerging biotechnologies are constantly undermining. Radical and methodological posthumanism, insofar as they develop from within a non-humanist framework, offer far better perspectives. But these frameworks also present significant limitations that need to be addressed. Methodological posthumanism does not carry through the

T. Sharon, *Human Nature in an Age of Biotechnology: The Case for Mediated Posthumanism*, Philosophy of Engineering and Technology 14, DOI 10.1007/978-94-007-7554-1_9, © Springer Science+Business Media Dordrecht 2014

theoretical ramifications its analyses have for subjectivity and needs to address the normative implications of the interwoven nature of humans and technologies. Radical posthumanism on the other hand, too often falls back onto a dichotomous framework that views emerging biotechnologies as embodying a schizophrenic, deconstructive potential that is captured and rechanneled onto fixed, conventional categories of nature and the human. This framework does not do enough to account for the intricate ways in which schizophrenic and paranoid trends coincide in understandings of nature and subjecthood in light of emerging biotechnologies today.

Mediated posthumanism aims to overcome the limitations of these non-humanist approaches. First, by applying a richer understanding of the concept of reterritorialization in the context of emerging biotechnologies – as the production of a new system of meaning in which deterritorialized elements connect in different ways, rather than a mere reconstitution of an older system of meaning. This makes it possible to account for categories and processes which should be identified as novel ones. In this sense, it is not a case of just identifying the de-naturalization vs. the re-naturalization of nature, genetic reductionism vs. genetic complexity, genetic determinism vs. genetic flexibility, depth vs. surface ontologies, etc., but of accounting for how these narratives exist simultaneously, in a "digital" sense in which contrary aspects can momentarily be brought into focus or can be minimalized while never cancelling each other out. Thus "re-naturalized" nature might remain a foundational category that normalizes and legitimizes, but its referent is constantly changing in accordance to who uses the term "nature" and in which context, not in a hegemonic sense, but in a more strategic, personalized sense, that allows new aspects of being human to be included into its realm. The notion of life, too, in the genetic context, is seen both as something that can be controlled and manipulated, but it also retains a metaphysical, ungraspable and unrepresentable aspect. And selfhood, while it is increasingly being brought into the realm of malleability and choice as it is redefined in terms of biology, of genetic dispositions and neurochemical balances, remains anchored in notions of authenticity and profound truths.

This process is more complex than a mere paranoid capture of the destabilizing effects of emerging biotechnologies via a clear reinforcement of the boundary lines between nature and culture, or a simply negative movement of reterritorialization that stifles the innovative and liberatory potential inherent in these technologies. This is not to say that conservative strategies that extend old concepts are not part of this dynamic, and that we must not remain vigilant about the paranoid tendencies that are often inherent to the use of emerging biotechnologies. It is to say that there is much more going on here: this understanding of the movements of de- and reterritorialization is helpful in grasping the underlying paradox of emerging biotechnologies, the idea that nature, or biology, is both given and given to control, that the natural or biological existence of things is both determined and that it can be intervened upon and remade.

Secondly, mediated posthumanism aims to bring together Foucault's ethical work on subject constitution with the notion of technological mediation, as a framework in which emerging biotechnologies can be construed as technologies of the self. Here ethics is a matter of stylizing those relationships to the powers, among them technology, that constitute the self, of developing a productive relationship to those relations. But this takes place within a non-humanist understanding of subjectivity as an effect of power relations, not as something that stands in contrast to or outside of them. Freedom is thus not about escaping the structures of power, the technological mediations, that underlie subject constitution, but about engaging in an active relationship with them in order to modify their impact in desirable ways. This mode of freedom does not center on the autonomous, rational moral agent that stands in opposition to a technological world from which it must be protected, in dystopic posthumanist terms, or which it must learn to manipulate in order to enhance that autonomy, in liberal posthumanist terms; nor is it completely determined or independent in relation to the technological powers that make up its world. Like the novel categories of "nature", "life" and "authentic selfhood" that are emerging in light of new biotechnologies, subjectivity here is also both given and given to control, determinative, or transformative, but not deterministic.

Drawing on these two theoretical frameworks and elements in radical and methodological posthumanism, mediated posthumanism also offers a means of identifying the positive and enriching effects of emerging biotechnologies on human experience. The notion of technological mediation implies that technology is a form of engagement with the world, and that every technological practice, every incorporation of technology into human experience, allows new forms of engagement to take place. These may be harmful, but they may also be enriching, and in any case this cannot be determined in advance.

It is this kind of openness to the novel forms of interaction with technologies that can allow us to identify genetic responsibility as a form of ethics of the self, that emerges from within the new obligations towards moral decision-making, self-actualization and responsibilities towards oneself and others that genetic risk management entails. It is true that the enactment of genetically responsible behavior has become something that is expected of all individuals, from the ill to the potentially at-risk, and that this responsibility, based on one's biological identity, intersects with other forms of self-government characteristic of neoliberal ideology. But as discussed, biological conceptions of personhood and genetic identity are rarely hegemonic. If we look closely we can perceive a number of behaviors that escape both the requirement to practice rational choice and autonomy, and the disciplinary concerns of critical approaches. Fears of a new biological determinism, or "geneticization", have not been realized. While individuals may increasingly draw upon biological narratives to articulate their subjectivity, references to genetics or neuroscience are rarely the only or even dominant narratives being used. Rather, these are often incorporated and combined with non-biological narratives in a creative and strategic mobilization of the biological for different ends in different contexts. Furthermore, the shift of health responsibilities from state to individual, a main concern of critics of health promotion, has led to resourceful practices on behalf of users that can hardly be described in terms of having or lacking autonomy, of individual choice or discipline. Thus the internet has become the site for a vast crowdsourcing of medical decisions where new forms of solidarity are created. And genetic knowledge is often appropriated by users in unexpected ways. All of these practices can be seen as creative processes of identity formation, where both technology and biology are given, but given as starting points, that mediate what it means to be human.